Circuit Design and Analysis

Other McGraw-Hill Reference Books of Interest

Handbooks

BENSON • *Audio Engineering Handbook*

BENSON • *Television Engineering Handbook*

COOMBS • *Printed Circuits Handbook*

DI GIACOMO • *Digital Bus Handbook*

DI GIACOMO • *VLSI Handbook*

FINK AND CHRISTIANSEN • *Electronics Engineers' Handbook*

HARPER • *Electronic Packaging and Interconnection Handbook*

HICKS • *Standard Handbook of Engineering Calculations*

INGLIS • *Electronic Communications Handbook*

JURAN AND GRYNA • *Juran's Quality Control Handbook*

KAUFMAN AND SEIDMAN • *Handbook of Electronics Calculations*

STOUT AND KAUFMAN • *Handbook of Operational Amplifier Circuit Design*

TUMA • *Engineering Mathematics Handbook*

WILLIAMS AND TAYLOR • *Electronic Filter Design Handbook*

Other

ANTOGNETTI • *Power Integrated Circuits*

ANTOGNETTI AND MASSOBRIO • *Semiconductor Device Modeling with SPICE*

BUCHANAN • *CMOS/TTL Digital Systems Design*

BUCHANAN • *BiCMOS/CMOS Systems Design*

BYERS • *Printed Circuit Board Design with Microcomputers*

ELLIOTT • *Integrated Circuits Fabrication Technology*

HECHT • *The Laser Guidebook*

MUN • *GaAs Integrated Circuits*

SILICONIX • *Designing with Field-Effect Transistors*

SZE • *VLSI Technology*

TSUI • *LSI/VLSI Testability Design*

WATERS • *Active Filter Design*

WOBSCHALL • *Circuit Design for Electronic Instrumentation*

WYATT • *Electro-Optical System Design*

Circuit Design and Analysis

Featuring C Routines

C. Britton Rorabaugh

McGraw-Hill, Inc.

New York St. Louis San Francisco Auckland Bogotá
Caracas Lisbon London Madrid Mexico Milan
Montreal New Delhi Paris San Juan São Paulo
Singapore Sydney Tokyo Toronto

Library of Congress Cataloging-in-Publication Data

Rorabaugh, Britt.
Circuit design and analysis featuring C routines / C. Britton Rorabaugh.
p. cm.
ISBN 0-07-053653-8
1. Electric filters, Digital—Design and construction—Data processing. 2. Electronic circuit design—Data processing. 3. Electronic circuit analysis—Data processing. 4. C (Computer program language) I. Title.
TK7872.F5R67 1992
621.381'5324—dc20 91-18931
CIP

1 2 3 4 5 6 7 8 9 0 DOC/DOC 9 7 6 5 4 3 2 1

P/N 053655-4
PART OF
ISBN 0-07-053653-8

The sponsoring editor for this book was Daniel A. Gonneau, the editing supervisor was Peggy Lamb, and the production supervisor was Suzanne W. Babeuf. This book was set in Century Schoolbook by Datapage.

Printed and bound by R. R. Donnelley & Sons Company.

LIMITS OF LIABILITY AND DISCLAIMER OF WARRANTY
The author and publisher have exercised care in preparing this book and the programs contained in it. They make no representation, however, that the programs are error-free or suitable for every application to which the reader may attempt to apply them. The author and publisher make no warranty of any kind, expressed or implied, including warranties of merchantability or fitness for a particular purpose, with regard to these programs or the documentation or theory contained in this book, all of which are provided "as is." The author and publisher shall not be liable for damage in connection with, or arising out of the furnishing, performance, or use of these programs or the associated descriptions or discussions.

Readers should test any program in their own systems and compare results with those presented in this book. They should then construct their own test programs to verify that they fully understand the requisite calling conventions and data formats for each of the programs. Then they should test the specific application thoroughly.

To Joyce, Geoff, and Amber

Contents

Preface xi

Chapter 1 Introduction and Background 1
1.1 Motivation 1
1.2 About the Programs in This Book 2

Chapter 2 Network Elements 5
2.1 Ideal Resistors 5
2.2 Independent Sources 8
2.3 Ideal Capacitors 9
2.4 Ideal Inductors 12
2.5 Transformers 14
2.6 Practical Components 15

Chapter 3 Useful Results from Network Theory 19
3.1 Network Configurations 19
3.2 Kirchhoff's Laws 27
3.3 Voltage Dividers 28
3.4 Superposition and Reciprocity 28
3.5 Norton and Thevenin Transformations 30

Chapter 4 Solid-State Circuit Elements 31
4.1 Additional Ideal Elements 31
4.2 Diodes 32
4.3 Transistor Models 35

Chapter 5 Circuit Specifications 39
5.1 Networks as Linear Systems 39
5.2 Characterization of Linear Systems 41
5.3 Transfer Functions 43
5.4 Manipulating Transfer Functions 48
5.5 Partial Fraction Expansion 51
5.6 Continued Fractions 53
5.7 Frequency Response 58

Chapter 6 Filter Specifications 61
6.1 Filter Fundamentals 61

6.2 Butterworth Filters 70
6.3 Chebyshev Filters 81
6.4 Bessel Filters 99

Chapter 7 Matrix Methods 107
7.1 Matrix Fundamentals 107
7.2 Cramer's Rule 110
7.3 Gaussian Elimination 110
7.4 Triangular Factorization 122

Chapter 8 Network Topology 129
8.1 Graphs 129
8.2 Incidence Matrix 132
8.3 Loop Matrix 132
8.4 Cutset Matrix 133

Chapter 9 Network Equations 135
9.1 Nodal Admittance Formulation 135
9.2 Construction Rules for the Nodal Admittance Formulation 140
9.3 Tableau Formulation 144
9.4 Augmented Nodal Formulation 147

Chapter 10 Synthesis of Passive Networks 163
10.1 Input Impedance 163
10.2 Foster Synthesis 164
10.3 Cauer Synthesis 169
10.4 Source-Terminated *LC* Ladder Networks 175
10.5 Load-Terminated *LC* Ladder Networks 178
10.6 Double-Terminated *LC* Ladder Networks 181

Chapter 11 Symbolic Network Functions 185
11.1 Signal Flow Graphs 185
11.2 Mason's Rule 191
11.3 Tree Enumeration Method 193
11.4 Parameter Extraction Method 197

Chapter 12 Sensitivity 199
12.1 Sensitivity Concepts 199
12.2 Incremental-Network Approach 200

Chapter 13 Design Studies 203
13.1 Approximation and Synthesis 203
13.2 Performance 205
13.3 Compensating for Lossy Inductors 209

Appendix A Global Definitions 211

Appendix B Prototypes for C Functions 213

Appendix C Functions for Complex Arithmetic 217

Appendix D Laplace Transform 223

Index 227

Preface

Now anyone involved in the design or modification of electronic circuits can take advantage of powerful, computer-aided circuit analysis and synthesis techniques which are presented in this book. All the methods presented are suitable for use on small personal computers, and some methods are simple enough to be used with pencil, paper, and calculator for smaller circuits. Where appropriate, the presentation of each computational method includes one or more C language programs which implement the method. Each program is given in the form of self-contained function which can be called from any main program of the reader's devising. All the functions employ a common system of data structures and indexing conventions so that the outputs of one function are directly usable as inputs by subsequent functions in the processing sequence. Included with the book is an IBM PC-compatible disk containing the source code for each program in the book. For the benefit of inexperienced C users, programming conventions and data-structure particulars are detailed in the text.

A review of circuit analysis fundamentals includes:

- Ideal network elements and their properties
- How practical "real-world" components differ from their ideal counterparts and how these differences can be accounted for in circuit analyses
- Circuit and network terminology
- Fundamental network theorems upon which all circuit analysis methods are based
- Circuit specifications

The heart of the book is devoted to practical computational methods which include:

- Several algorithms for solution of matrix equations
- Computer representation of network topologies
- Various approaches for construction of a network's system equation
- Algorithms for synthesis of passive networks
- Generation of symbolic network functions
- Sensitivity analysis

Britt Rorabaugh

Circuit Design and Analysis

Chapter

1

Introduction and Background

This book presents computational methods for the synthesis and analysis of electronic circuits. All the methods presented are suitable for use on small personal computers, and some are simple enough to be used with pencil, paper, and calculator for smaller circuits.

1.1 Motivation

Engineers are divided into two schools of thought, concerning computational methods for circuit analysis. One group feels that since "canned" circuit analysis programs such as SPICE are readily available, there is no need for the average engineer to know what goes on "inside the can." The other group feels that while programs such as SPICE are useful tools for circuit analysis, they by no means obviate the need for a working knowledge of computational techniques. This second view is supported by the following arguments:

- The generality of canned analysis programs is obtained in exchange for relatively large memory requirements and relatively slow execution speeds. A dedicated analysis program configured from the modules presented in this book will have greatly reduced memory requirements and will execute much faster than general-purpose canned programs.
- Some powerful techniques for circuit optimization require access to intermediate results not usually available from canned programs. Furthermore, the iterative nature of these optimization techniques makes the use of dedicated analysis programs almost mandatory in order to avoid prohibitively long execution times.
- It is possible to form complete "end-to-end" design programs by coupling the synthesis and analysis methods presented in this book. This is extremely difficult, if not impossible, to do with canned analysis programs.

The techniques presented in Chap. 6 are used to generate the appropriate filter function for a desired set of performance specifications. The synthesis techniques presented in Chap. 10 can then be used to find a network of ideal passive elements which realizes the required filter function. The sensitivity techniques of Chap. 12 are then used in conjunction with the analysis techniques of Chaps. 11, 9, 8, 7, and 5 to determine the actual filter performance achievable when practical components are substituted for ideal components. Finally, the techniques of Sec. 13.3 can be used to compensate for the effects of lossy inductors in *RLC* circuits. Sometimes, however, the indicated performance will still be inadequate. In such cases, the program must return to the data of Chap. 6 to select a filter function which offers improved performance in the deficient areas.

The algorithms and programs presented in this book will prove to be very useful to engineers, students, and hobbyists engaged in circuit design and analysis.

1.2 About the Programs in This Book

All the programs in this book were written and tested using Think C™ for the Apple Macintosh computer. A conscientious effort was made to limit the programs to the ANSI standard subset of Think C™ and to avoid any machine dependencies. Potential efficiencies were sacrificed for the sake of portability and tutorial clarity. However, a few specific items need to be pointed out.

Global definitions and ANSI prototyping

Constants used by several different functions are collected into a single include file called "globDefs.h"—a listing of this file is provided in App. A. The "new" style of ANSI prototyping was used throughout all the software generated for this book. All the pertinent prototypes are collected in a file called "protos.h" which is given in App. B.

File naming

Nice long file names such as "BuildNodalAdmittanceMatrix.c" are allowed on the Macintosh, but on MS-DOS machines file names are limited to eight characters plus a three-character extension. Except for the two header files mentioned above, all the files on the accompanying disk have names that are keyed to the Listing numbers printed in the text. For example, Listing 7.2 is contained in file Lis_7_2.c and Listing 13.1 is contained in file Lis_13_1.c.

Type definitions

I found it convenient to define a new type real that is the same as double. For use on machines with limited memory, real could be redefined as float to save memory, but accuracy and numerical stability could suffer. Being a long-time FORTRAN user, I also found it convenient to create a logical type rather than the usual boolean. The lack of intrinsic complex types in C was overcome via a complex structure definition and a set of complex arithmetic functions detailed in App. C.

Chapter

2

Network Elements

This chapter presents a review of ideal network elements and their properties, along with a discussion of how practical "real-world" components differ from their ideal counterparts and how these differences can be accounted for in circuit analyses.

2.1 Ideal Resistors

An ideal resistor is a two-terminal device that can be characterized by a functional relationship between the voltage across the device and the current through the device. Either one or both of the following will be true:

$$v = f_R(i) \qquad i = f_G(v)$$

In the simplest case, the relationship reduces to Ohm's law, which is given by

$$v = Ri$$

where the *resistance* R is time-invariant and independent of both v and i. Such a resistor is said to be *linear* and is characterized by a straight line through the origin of the voltage-versus-current plane, as shown in Fig. 2.1. The slope of the line is equal to R. It may be tempting to refer to any resistor having a straight-line v-i characteristic as linear, but the linearity property presented in Sec. 4.1 will not be satisfied unless the line passes through the origin. In general, resistance can be a function of voltage, current, and time

$$R = f_R(v, i, t)$$

and nonlinear V-I characteristics are possible. If the relationship between voltage and current is a function of current, as shown in Fig. 2.2, the resistor is said to be *current-controlled*; and when the relationship is a function of voltage, as shown in Fig. 2.3, the resistor is said to be *voltage-controlled*. A strictly monotonic v-i characteristic (such as for constant R) can be

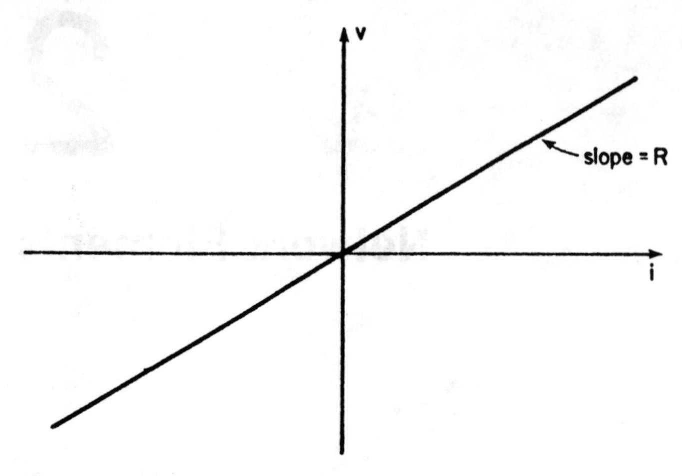

Figure 2.1 Typical v-i characteristic for a linear resistor.

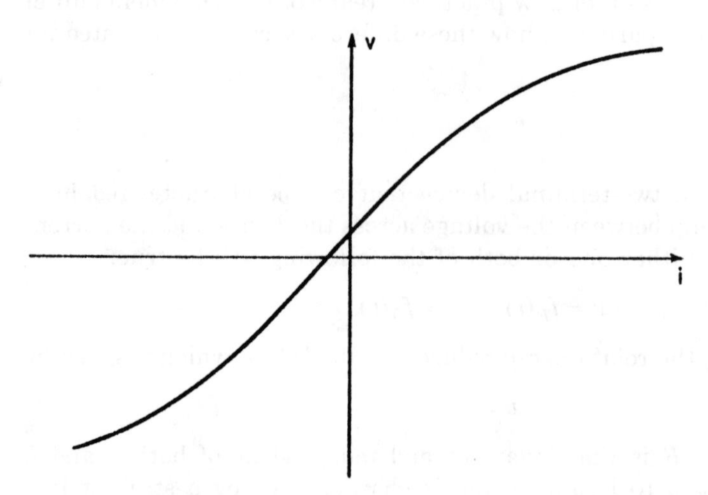

Figure 2.2 Typical v-i characteristic for a current-controlled resistor.

expressed as either a function of current or a function of voltage, and the corresponding resistor is both current-controlled and voltage-controlled. Nonlinear resistors may seem somewhat of a perversity, but they appear quite often in equivalent circuit models for semiconductor devices. The usual schematic symbols for linear resistors and nonlinear resistors are shown in Fig. 2.4.

When the resistance function is invertible (i.e., the resistor is both current-controlled and voltage-controlled), the inverse is called *conductance* and is usually denoted by the letter G. The straight-line characteristic of a linear resistor will have a slope of G if plotted in the current-versus-voltage plane instead of the voltage-versus-current plane. When voltage is measured in

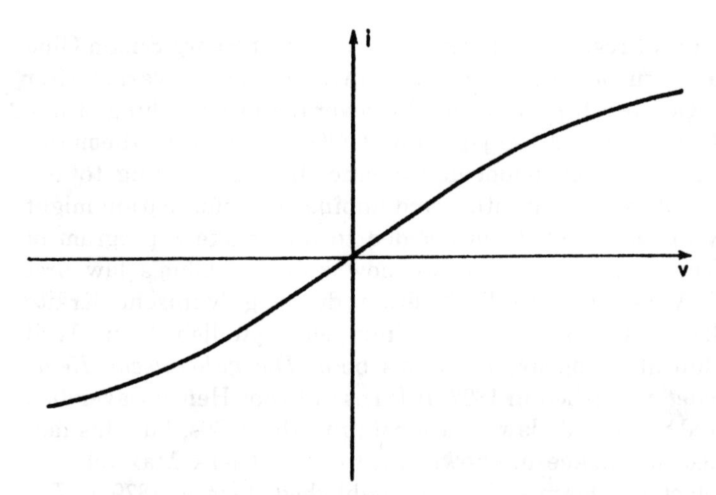

Figure 2.3 Typical i-v characteristic for a voltage-controlled resistor.

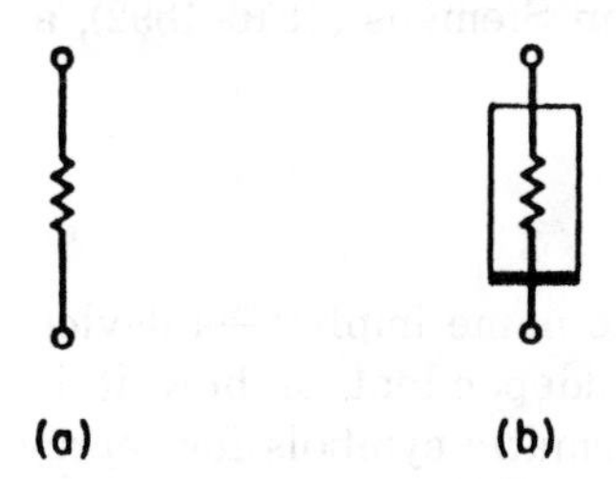

Figure 2.4 Schematic symbols for (*a*) linear resistor and (*b*) nonlinear resistor.

volts and current is measured in amperes, resistance is in *ohms*. The uppercase Greek letter omega (Ω) is often used to denote ohms. The traditional unit of conductance is the *mho*, denoted by an inverted omega (℧), but the official SI unit of conductance is the *siemen*, denoted by an uppercase roman S.

In the limit, a short circuit can have an arbitrary current flowing with no corresponding voltage developed across the terminals. This can be viewed as a current-controlled resistor having a constant resistance of zero over all currents. Similarly, an open circuit can have an arbitrary voltage with no corresponding current and can thus be viewed as a voltage-controlled resistor having a constant conductance of zero over all voltages.

Power dissipation. The power dissipated in a resistor is equal to the product of voltage and current

$$P = VI = I^2R = V^2R$$

When V is in volts (V) and I is in amperes (A), P is in *watts* (W).

Historical note. The unit of resistance is named in honor of Georg Simon Ohm (1789–1854) who was born on March 16, 1789, in Erlangen, Bavaria. After receiving a Ph.D. in October 1811, followed by several short teaching stints, in September 1817 Ohm accepted the position of *Oberlehner* of mathematics and physics at the Jesuit Gymnasium at Cologne. In 1825, feeling totally dissatisfied with his professional situation and hoping that publication might lead to a university appointment, Ohm decided to undertake a program of original research. The relationship that we now know as Ohm's law first appeared in Ohm's "Versuch einer Theorie der durch galvanische Kräfte hervorgebracten elecktroskopischen Erscheinungen," published in April 1826. The relationship also appears in Ohm's book *Die galvanische Kette, mathematisch bearbeitet*, published in 1827. It turns out that Henry Cavendish (1731–1810) discovered the Ohm's law relationship in the 1770s, but this fact remained unpublished and hence unknown until James Clerk Maxwell collected Cavendish's electrical manuscripts and published them in 1879 as *The Electrical Researches of the Honourable Henry Cavendish*. The SI unit of conductance is named in honor of Ernst Werner von Siemens (1816–1892), a German scientist and inventor.

2.2 Independent Sources

An *independent voltage source* (IVS) is just what the name implies—a device which sources or supplies a particular voltage, independent of how it is connected to other circuit elements. The usual schematic symbols for representing independent voltage sources are shown in Fig. 2.5. As shown in Fig. 2.6, the v-i characteristic for an IVS is a horizontal line. For an IVS with $v = 0$, the characteristic coincides with the i axis as does the characteristic for a short circuit or a current-controlled resistor with $R = 0$. Thus to remove the *effects* of an IVS from a circuit under analysis, the IVS can be replaced with a short circuit. Although the IVS is an idealization that does not exist, many practical voltage sources can be modeled as an IVS in series with an ideal resistor. For many types of chemical batteries, the value of this resistor is specified as the battery's internal resistance. The lower this resistance is, the more closely the voltage source approaches the ideal.

An *independent current source* (ICS) is a device that supplies a particular current independent of how the device is connected to other circuit elements.

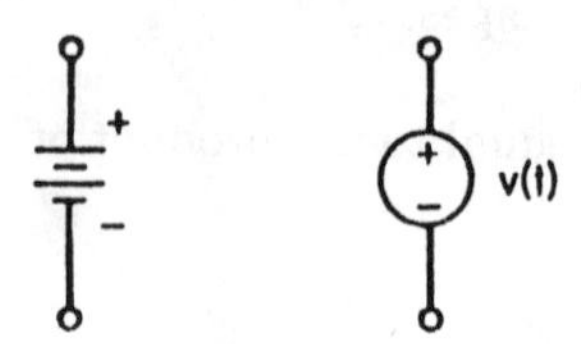

Figure 2.5 Schematic symbols for independent voltage sources.

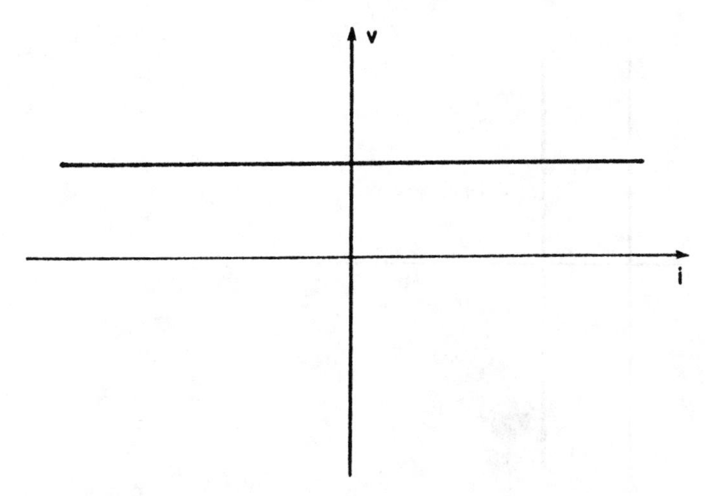

Figure 2.6 Typical v-i characteristic for an independent voltage source.

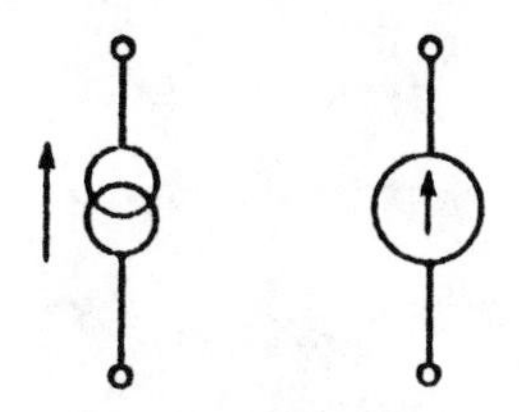

Figure 2.7 Schematic symbols for independent current sources.

The usual schematic symbols for representing independent current sources are shown in Fig. 2.7. As shown in Fig. 2.8, the v-i characteristic for an ICS is a vertical line. For an ICS with $i = 0$, the characteristic coincides with the v axis as does the characteristic for an open circuit or a voltage-controlled resistor with $G = 0$. Thus to remove the *effects* of an ICS from a circuit under analysis, the ICS can be replaced with an open circuit.

2.3 Ideal Capacitors

In its simplest form, a capacitor can be modeled as two conductive plates separated by a layer of nonconductive material, as shown in Fig. 2.9. The layer of nonconductive material is called the *dielectric*. A voltage applied to the capacitor's terminals will cause an accumulation of positive charges on one plate and an accumulation of negative charges on the other plate. The total charge is a function of the applied voltage

$$q = f(v) \tag{2.1}$$

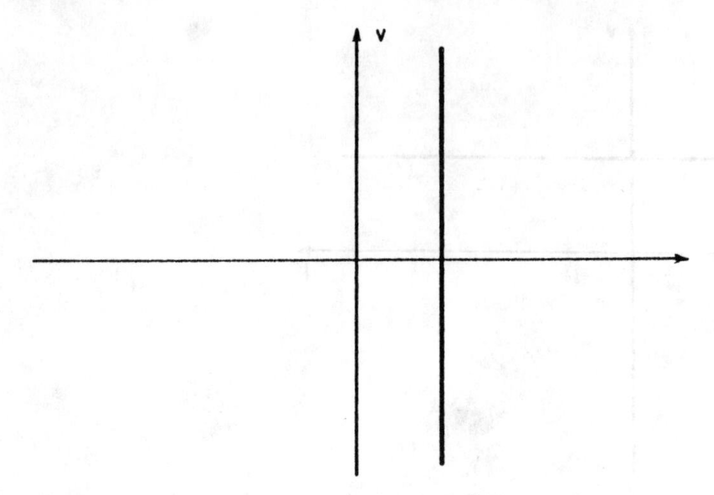

Figure 2.8 Typical v-i characteristic for an independent current source.

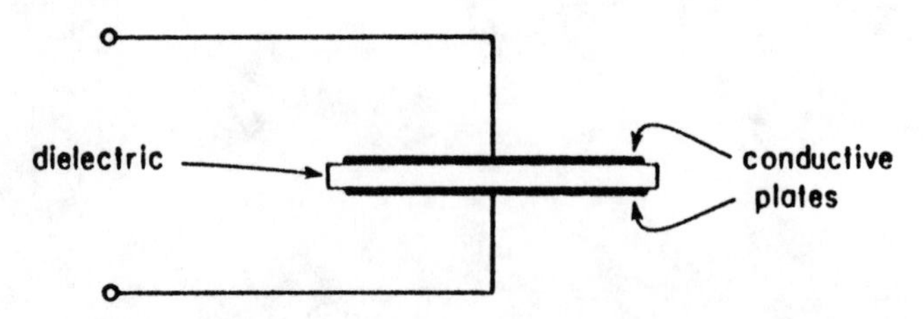

Figure 2.9 A simple capacitor.

For a linear time-invariant capacitor, (2.1) reduces to

$$q = Cv \tag{2.2}$$

where C denotes *capacitance*. When charge is measured in coulombs (C) and voltage is measured in volts, capacitance is in *farads* (F). The current through the capacitor is the time derivative of charge

$$i(t) = \frac{d}{dt} q(t) = C \frac{d}{dt} v(t) \tag{2.3}$$

Integration yields

$$\begin{aligned} v(t) &= \frac{1}{C} \int_{-\infty}^{t} i(\tau)\, d\tau \\ &= \frac{q_0}{C} + \frac{1}{C} \int_{0}^{t} i(\tau)\, d\tau \\ &= V_0 + \frac{1}{C} \int_{0}^{t} i(\tau)\, d\tau \end{aligned} \tag{2.4}$$

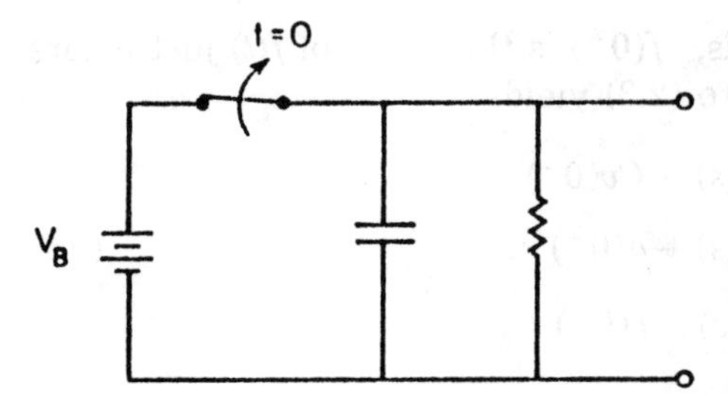

Figure 2.10 Circuit for charging a capacitor until time $t = 0$.

where V_0 is the voltage appearing across the capacitor at time $t = 0$ and q_0 is the charge on the capacitor at time $t = 0$. Note that V_0 is not the external voltage applied to the capacitor, but rather the terminal voltage due to the stored charge. In many cases, however, the "stored" voltage and externally applied voltage are the same. For the circuit shown in Fig. 2.10, the initial capacitor voltage V_0 at time $t = 0$ is equal to the battery voltage. However, for the circuit shown in Fig. 2.11, one switch opens at time $t = -1$, isolating the capacitor and allowing the accumulated charge to remain. Thus, at time $t = 0$, the initial capacitor voltage is $V_0 = V_B$ even though the external voltage source was previously disconnected.

Using equations such as (2.3) and (2.4) to analyze circuits will generally result in systems of simultaneous integrodifferential equations that need to be solved. A (usually) easier-to-solve system of algebraic equations can be obtained by taking the Laplace transform of (2.3). From Table D.2 in App. D we make use of property #4

$$\mathscr{L}\left[\frac{d}{dt}f(t)\right] = sF(s) - f(0^-) \tag{2.5}$$

where $f(0^-)$ denotes the *left-hand limit*, or the limit of $f(x)$ as x approaches zero from the left

$$f(0^-) \triangleq \lim_{t \to 0^-} f(t)$$

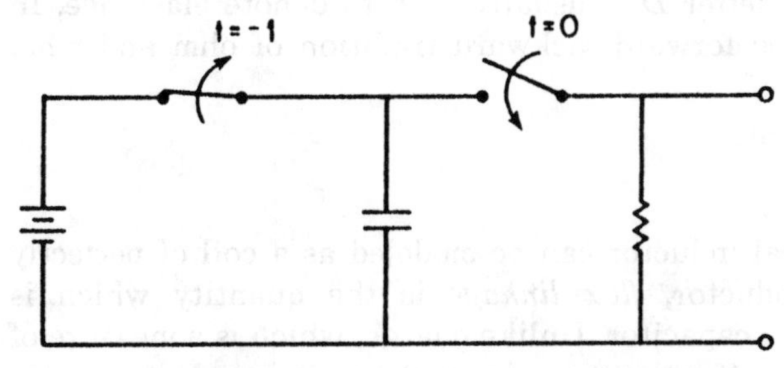

Figure 2.11 Circuit for charging a capacitor until time $t = -1$ and then beginning discharge at time $t = 0$.

[A more informal way of saying this is, "$f(0^-)$ is the value of $f(t)$ just before time $t = 0$."] Application of Eq. (2.5) to (2.3) yields

$$I(s) = sCV(s) - Cv(0^-)$$

$$= sCV(s) - q(0^-) \tag{2.6}$$

$$V(s) = \frac{I(s)}{sC} + \frac{v(0^-)}{s}$$

$$= \frac{I(s) + q(0^-)}{sC} \tag{2.7}$$

In subsequent chapters, Eqs. (2.3), (2.6), and (2.7) are referred to as *constitutive equations* for a linear time-invariant capacitor. To use Eqs. (2.6) and (2.7), it is necessary to first obtain the Laplace transform of all voltages and currents that are given as functions of time. The results will be in the s domain and will need to be inverse-transformed to obtain time-domain voltages or currents.

The quantity $(2\pi fC)^{-1}$ is called the *capacitive reactance* and is usually denoted by X_c. The impedance of a resistor and capacitor in series is given by

$$Z = \sqrt{R^2 + X_c^2}$$

The quantity $2\pi fC$ is called the *capacitive susceptance* and is usually denoted by B_c. A parallel combination of a conductance and a capacitance will have a total admittance given by

$$Y = \sqrt{G^2 + B_c^2}$$

Note that an instantaneous step change in the voltage across a capacitor implies an infinite derivative at the instant of change and hence, as a consequence of (2.3), an infinite current (assuming that C is time-invariant). Since infinite currents are not physically realizable, neither are instantaneous changes in voltage across a capacitor.

The unit of capacitance is named in honor of Michael Faraday (1791–1867), an English physicist. The defined (but rarely used) reciprocal of capacitance is called the *elastance*. The letter D is usually used to denote elastance. In keeping with the established forward-backward tradition of ohm and mho, the unit of elastance is *daraf*.

2.4 Ideal Inductors

In its simplest form, an ideal inductor can be modeled as a coil of perfectly conductive wire. In an inductor, *flux linkage* is the quantity which is analogous to the charge on a capacitor. Unlike charge, which is a measure of the number of extra electrons (for negative charges) or missing electrons (for positive charges), flux linkage does not have such straightforward physical significance. The unit of flux linkage is the intellectually unsatisfying *weber-turn* (Wb·turn). Furthermore, the notational issues begin to get just a bit sticky. Many texts use Φ to denote the total flux and then assume that all the turns of an N-turn coil link with this flux, resulting in a flux linkage of $N\Phi$. Other texts make the more realistic but messier assumption that not all turns link with the same flux. This leads to expressions such as

$$\psi = \sum_i N_i \Phi_i \tag{2.8}$$

where ψ = flux linkages
N_i = number of turns which link flux of Φ_i

Rather than trying to compute the flux linkages from the flux, still other texts simply denote the flux linkage as ψ and move on. Since this is not a book on field theory or inductor design, we take this last approach. In any event, the flux linkage ψ is a function of the current through the inductor

$$\psi = f(i) \tag{2.9}$$

For a linear time-invariant inductor, this reduces to

$$\psi = Li \tag{2.10}$$

where L is the inductance. When flux linkage is measured in weber-turns and current is measured in amperes (A), then inductance will be in *henrys* (H).

The voltage across an inductor is the time derivative of the flux linkage

$$v(t) = \frac{d}{dt}\psi(t) = L\frac{d}{dt}i(t) \tag{2.11}$$

Integration yields

$$\begin{aligned} i(t) &= \frac{1}{L}\int_{-\infty}^{t} v(\tau)\,d\tau \\ &= \frac{1}{L}\int_{-\infty}^{0} v(\tau)\,d\tau + \frac{1}{L}\int_{0}^{t} v(\tau)\,d\tau \\ &= \frac{\psi_0}{L} + \frac{1}{L}\int_{0}^{t} v(\tau)\,d\tau \\ &= I_0 + \frac{1}{L}\int_{0}^{t} v(\tau)\,d\tau \end{aligned} \tag{2.12}$$

where I_0 is the current at time $t = 0$ and ψ_0 is the flux linkage at time $t = 0$. Just as for the case of capacitors, we can obtain easier-to-use constitutive equations for the inductor by taking the Laplace transform of (2.11) to obtain

$$V(s) = sLI(s) - Li(0^-)$$
$$= sLI(s) - \psi(0^-) \tag{2.13}$$

$$I(s) = \frac{V(s)}{sL} + \frac{i(0^-)}{s}$$
$$= \frac{V(s) + \psi(0^-)}{sL} \tag{2.14}$$

The quantity $2\pi fL$ is called the *inductive reactance* and is usually denoted by X_L. The impedance of a resistor and inductor in series is given by

$$Z = \sqrt{R^2 + X_L^2}$$

The quantity $(2\pi fL)^{-1}$ is called the *inductive susceptance* and is usually denoted by B_L. A parallel combination of a conductance and an inductor will have a total admittance given by

$$Y = \sqrt{G^2 + B_L^2}$$

Note that an instantaneous step change in the current through an inductor implies an infinite derivative at the instant of change and hence, as a consequence of (2.11), an infinite voltage (assuming that L is time-invariant). Since infinite voltages are not physically realizable, neither are instantaneous changes in current through an inductor.

The unit of inductance is named in honor of Joseph Henry (1797–1878), a U.S. scientist. I've never seen a unit for inverse inductance, but how about *yrneh*?

2.5 Transformers

A transformer consists of two inductors that are coupled to each other via *mutual inductance*. The defining equations for a transformer are

$$V_1 = sL_1 I_1 \pm sMI_2$$

$$V_2 = sL_2 I_2 \pm sMI_1$$

where L_1 = inductance of coil 1
L_2 = inductance of coil 2
M = mutual inductance

A transformer is depicted schematically as shown in Fig. 2.12. The dots are used to indicate the relative phasing of the two coils. If both dots are at the

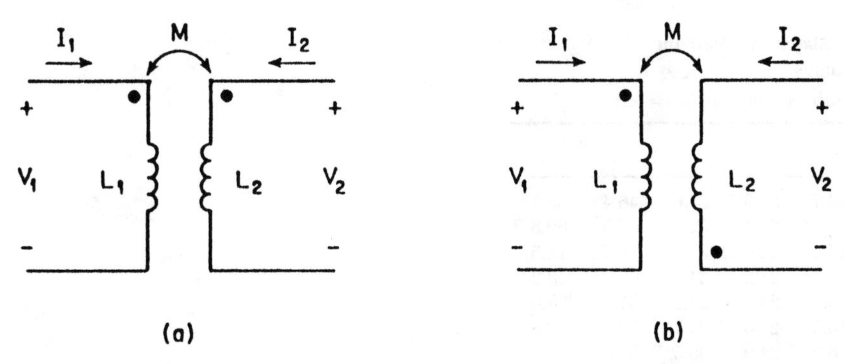

Figure 2.12 Schematic diagram of a transformer for (*a*) positive M and (*b*) negative M.

same end of their respective coils, the mutual inductance is positive. If the dots are at opposite ends, the mutual inductance is negative.

2.6 Practical Components

It is a fact of life that real-world electronic components have properties that often differ significantly from the properties of ideal elements discussed in previous sections. Practical resistors can come fairly close to the ideal, but practical inductors almost never do. Capacitors fall somewhere in the middle—sometimes close to ideal, sometimes not.

Resistors

Resistors come in a variety of types—carbon-composition, carbon-film, metal-film, and wirewound. Carbon-composition resistors are available in tolerances of ± 10 or ± 5 percent. Standard values for 10 and 5 percent resistors are listed in Tables 2.1 and 2.2. Carbon-film resistors are available in ± 5 and ± 1 percent tolerance ratings. Metal-film resistors are available in a "standard" ± 1 percent tolerance rating, but tolerances of ± 0.5, ± 0.25, and ± 0.1 percent can be obtained. Standard values for 1 percent resistors are listed in Table 2.3.

TABLE 2.1 Standard 10 Percent Resistor Values

standard values $= m \times 10^n$ $n = -1, 0, 1, \ldots, 7$

m			
10	18	33	56
12	22	39	68
15	27	47	82

TABLE 2.2 Standard 5 Percent Resistor Values

standard values $= m \times 10^n$ $n = -1, 0, 1, \ldots, 7$

m			
10	18	33	56
11	20	36	62
12	22	39	68
13	24	43	75
15	27	47	82
16	30	51	91

TABLE 2.3 Standard 1 Percent Resistor Values

standard values $= m \times 10^n$ $n = 0, 1, \ldots, 7$

m					
10.0	14.7	21.5	31.6	46.4	68.1
10.2	15.0	22.1	32.4	47.5	69.8
10.5	15.4	22.6	33.2	48.7	71.5
10.7	15.8	23.2	34.0	49.9	73.2
11.0	16.2	23.7	34.8	51.1	75.0
11.3	16.5	24.3	35.7	52.3	76.8
11.5	16.9	24.9	36.5	53.6	78.7
11.8	17.4	25.5	37.4	54.9	80.6
12.1	17.8	26.1	38.3	56.2	82.5
12.4	18.2	26.7	39.2	57.6	84.5
12.7	18.7	27.4	40.2	59.0	86.6
13.0	19.1	28.0	41.2	60.4	88.7
13.3	19.6	28.7	42.2	61.9	90.9
13.7	20.0	29.4	43.2	63.4	93.1
14.0	20.5	30.1	44.2	64.9	95.3
14.3	21.0	30.9	45.3	66.5	97.6

In critical applications a variable resistor can be used to adjust a resistance to any required value. Standard values are $n \times 10^m$, where $n = 1$, 2, or 5 and $m = 1$, 2, 3, 4, or 5.

Capacitors

In addition to capacitance, a practical capacitor generally exhibits an amount of series inductance, series resistance, and parallel resistance. Figure 2.13 shows an equivalent circuit for representing a practical capacitor in terms of ideal components. Standard capacitor values are of the form $m \times 10^n$ picofarads (pF), where m is one of the values for 5 percent resistors listed in Table 2.2.

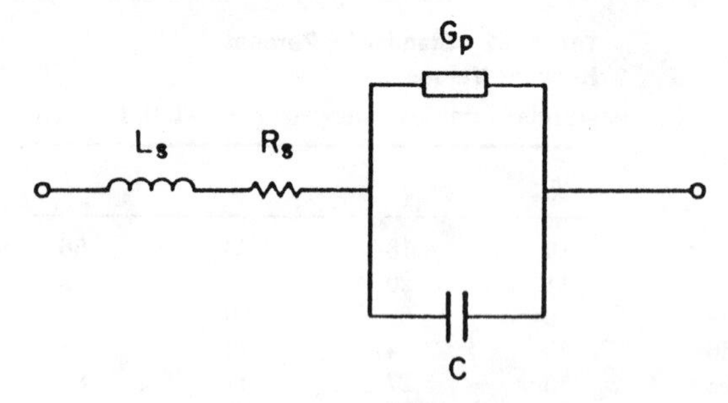

Figure 2.13 Equivalent circuit for a practical capacitor.

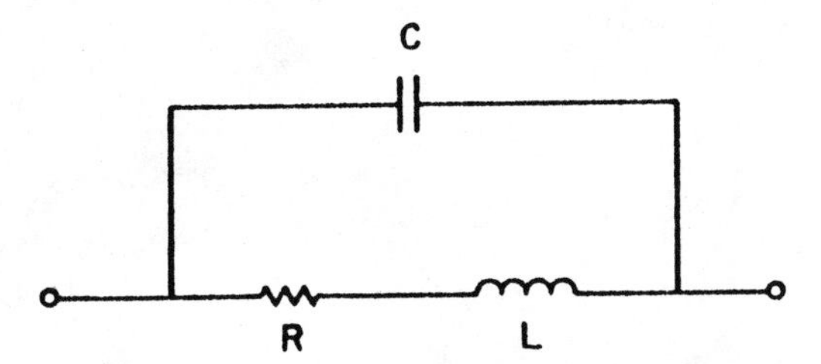

Figure 2.14 Equivalent circuit for a practical inductor.

Film capacitors are available with dielectrics of mylar, polycarbonate, polystyrene, and polypropylene, with mylar being the cheapest and polystyrene the best. Tolerances of 20, 10, 5, 2.5, and 1 percent are available. Film capacitors are usually limited to the range of 1000 pF to a few microfarads (μF).

Mica capacitors are available in values ranging from a few picofarads to about 0.09 μF, although the values in the high end of this range get to be very expensive. Ceramic capacitors are available in values ranging from 0.5 pF to 2.2 μF. They are inexpensive, but generally available only in tolerances of 10 or 20 percent. Also available are ceramic trimmer capacitors that adjust from nearly zero up to 50 or 100 pF.

Inductors

In addition to inductance, a practical inductor exhibits series resistance and parallel capacitance. Figure 2.14 shows an equivalent circuit for representing a practical inductor in terms of ideal components.

Chapter

3

Useful Results from Network Theory

This chapter defines basic circuit and network terminology and presents the fundamental network theorems upon which all circuit analysis methods are based.

3.1 Network Configurations

In this section we discuss the notions of *nodes* and *branches* as they apply to circuit theory. The schematic of a typical circuit is shown in Fig. 3.1. The points labeled *a*, *b*, and *c* are electrically the same point since they are all depicted as being directly connected by a perfect conductor. Likewise, the points labeled *g*, *h*, *k*, *m*, *n*, and *p* are all electrically a single point. We could, if we wished, take the schematic of Fig. 3.1 and redraw it as shown in Fig. 3.2. Examination of this figure reveals that points *a*, *b*, and *c* have been collapsed into a single point labeled 1; points *g*, *h*, *k*, *m*, *n*, and *p* have been collapsed

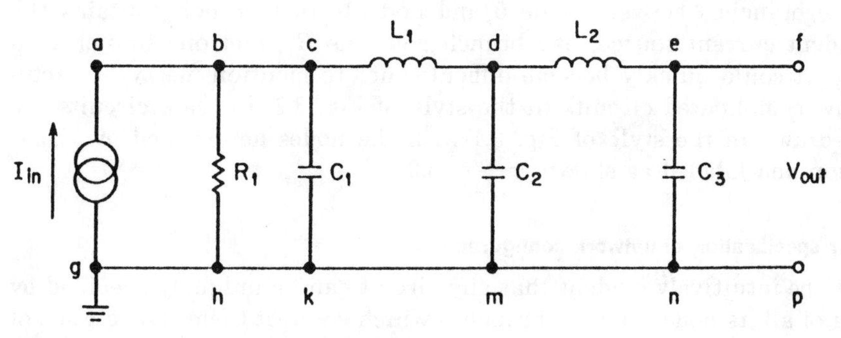

Figure 3.1 Schematic of a typical circuit.

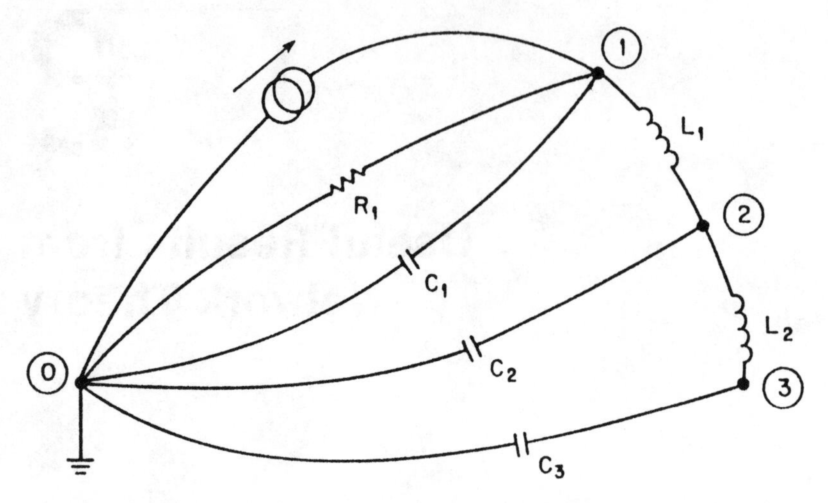

Figure 3.2 Schematic of Fig. 3.1 redrawn to clarify topology.

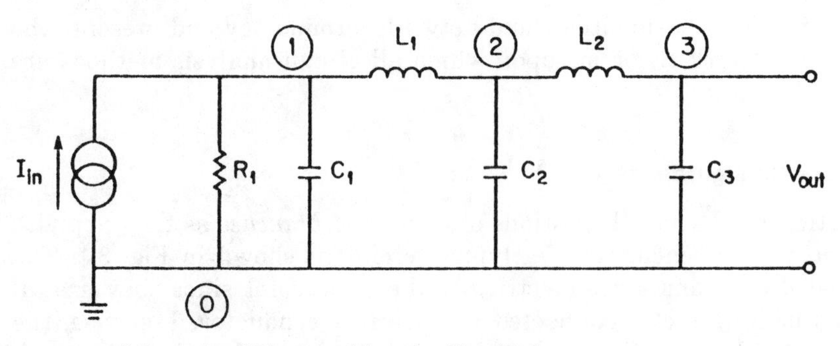

Figure 3.3 Schematic of Fig. 3.1 with nodes identified.

into a single point labeled 0; and points *e* and *f* have been collapsed into a single point labeled 3. Each of the points labeled in Fig. 3.2 is called a *node* of the circuit. Each pair of nodes is connected by one or more *branches*. There are three branches between node 0 and node 1—one branch contains the independent current source, one branch contains R_1, and one branch contains C_1. It could quickly become difficult—not to mention messy—to actually draw complicated circuits in the style of Fig. 3.2. Instead, circuits are usually drawn in the style of Fig. 3.1 with the nodes determined by visual inspection and labeled as shown in Fig. 3.3.

Computer specification of network configurations

It should be intuitively evident that any circuit can be uniquely specified by a listing of all its nodes and the branches which connect them. The circuit of Fig. 3.3 can be specified by the nodes and branches listed in Table 3.1. A

TABLE 3.1 Nodes and Branches for the Circuit Shown in Fig. 3.3

Type of element	Value	From node:	To node:
Current source	I_{in}	0	1
Resistor	R_1	0	1
Capacitor	C_1	0	1
Capacitor	C_2	0	2
Capacitor	C_3	0	3
Inductor	L_1	1	2
Inductor	L_2	2	3

human is usually happier with a value such as 0.1176 μF, but for the computer, 1.176×10^{-7} F is more convenient. Listing 3.1 contains a C routine **GetBranchData( )** that reads branch data records in "human format" and converts the data to "computer format." This routine will be used in conjunction with routines presented in Chap. 9 to automatically generate systems of network equations. Listing 3.1 also contains the functions **DecodeUnits( )** and **DecodeOnePort( )** which are used exclusively by **GetBranchData( )**. An annotated specimen of input data for **GetBranchData( )** is shown in Fig. 3.4.

```
/**************************************************/
/*                                                */
/*   Listing 3.1                                  */
/*                                                */
/*   GetBranchData()                              */
/*                                                */
/**************************************************/

#include "globDefs.h"
#include "protos.h"
#include <string.h>

extern FILE *inFile;
extern FILE *outFile;

int GetBranchData(
                    enum componentTypes *component,
                    char labels[][10],
                    real values[],
                    int *hNode,
```

```
                              int *iNode,
                              int *jNode,
                              int *kNode,
                              logical *done,
                              char depVar[],
                              char depVar2[])

{
char token[12][10];
char name[10], units[10];
real rawValue1, rawValue2;
enum unitTypes baseUnits;
int unitsIndex;
int returnCode;
real multiplier;
int i,j;
int testVal;
char cc;
/*----------------------------------------*/
/*  initialize the working storage        */
for(i=0;i<10;i++)
    {
    for(j=0;j<10;j++) token[i][j] = 0;
    }
*done=FALSE;
/*------------------------------------------------*/
/*  read one input record and separate the tokens  */

for(i=0; i<10; i++)
    {
    for(j=0; j<10; j++)
        {
        cc = fgetc(inFile);
        if((cc == EOF) || (cc == STOP_CHAR))
            {
            *done=TRUE;
            break;
            }
        if(cc == SPACE) break;
        if(cc == EOL) break;
        token[i][j] = cc;
        }
    fprintf(outFile,"%s ",token[i]);
    if((cc == EOL) || (cc == EOF)) break;
    }
    fprintf(outFile,"\n");
```

```
/*-------------------------------------*/
/*  decode the token contents          */

*component = undefined;
strcpy( name, token[0]);
if( !strcmp(name,"R\0")) *component = resistor;
if( !strcmp(name,"C\0")) *component = capacitor;
if( !strcmp(name,"L\0")) *component = inductor;
if( !strcmp(name,"G\0")) *component = conductance;
if( !strcmp(name,"I\0")) *component = currentSource;
if( !strcmp(name,"V\0")) *component = voltageSource;
if( !strcmp(name,"VCVS\0")) *component = VCVS;
if( !strcmp(name,"CCVS\0")) *component = CCVS;
if( !strcmp(name,"CCCS\0")) *component = CCCS;
if( !strcmp(name,"VCCS\0")) *component = VCCS;
if( !strcmp(name,"xfmr\0")) *component = transformer;
if( !strcmp(name,"OpAmp\0")) *component = OpAmp;
if(*component == undefined) return(-1);

/*-----------------------------------------------------------*/
/*  do detailed processing specific to each component type  */
values[1] = 0.0;
values[2] = 0.0;
*hNode = -1;
*iNode = -1;
returnCode = 0;
switch (*component)
    {
    case currentSource:
        if( DecodeOnePort(token,amps,values,labels,jNode,kNode) )
                return(-11);
        break;
    case voltageSource:
        if( DecodeOnePort(token,volts,values,labels,jNode,kNode) )
                return(-12);
        break;
    case resistor:
        if( DecodeOnePort(token,ohms,values,labels,jNode,kNode) )
                return(-13);
        strcpy( depVar, &(token[6][0]));
        break;
    case conductance:
        if( DecodeOnePort(token,mhos,values,labels,jNode,kNode) )
                return(-14);
```

```
        break;
    case capacitor:
        if( DecodeOnePort(token,farads,values,labels,jNode,kNode) )
                return(-15);
        break;
    case inductor:
        if( DecodeOnePort(token,henry,values,labels,jNode,kNode) )
                return(-16);
        strcpy( depVar, &(token[6][0]));
        break;
    case CCVS:
        strcpy( depVar2, &(token[8][0]) );
    case VCCS:
    case CCCS:
    case VCVS:
        strcpy(labels[0],token[1]);
        sscanf( &(token[2][0]), "%lg", &(values[0]) );
        sscanf( &(token[3][0]), "%d", hNode);
        sscanf( &(token[4][0]), "%d", iNode);
        sscanf( &(token[5][0]), "%d", jNode);
        sscanf( &(token[6][0]), "%d", kNode);
        strcpy( depVar, &(token[7][0]) );
        break;
    case transformer:
        strcpy( units, token[1]);
        DecodeUnits( units, &baseUnits, &multiplier);
        sscanf( &(token[3][0]), "%lg", &rawValue1);
        sscanf( &(token[5][0]), "%lg", &rawValue2);
        values[0] = multiplier * rawValue1;
        values[1] = multiplier * rawValue2;
        sscanf( &(token[7][0]), "%lg", &(values[2]) );
        strcpy(labels[0],token[2]);
        strcpy(labels[1],token[4]);
        strcpy(labels[2],token[6]);
        sscanf( &(token[8][0]), "%d", hNode);
        sscanf( &(token[9][0]), "%d", iNode);
        sscanf( &(token[10][0]), "%d", jNode);
        sscanf( &(token[11][0]), "%d", kNode);
        strcpy( depVar, &(token[9][0]) );
        strcpy( depVar2, &(token[10][0]) );
        break;
    case OpAmp:
        sscanf( &(token[1][0]), "%lg", &(values[0]) );
        sscanf( &(token[2][0]), "%d", hNode);
        sscanf( &(token[3][0]), "%d", iNode);
```

```
            sscanf( &(token[4][0]), "%d", jNode);
            sscanf( &(token[5][0]), "%d", kNode);
            strcpy( depVar, &(token[6][0]) );
            break;
        default:
            returnCode = 1;
        }   /*  end of switch  */

return(returnCode);
}
/****************************************/
/*                                      */
/*  DecodeUnits()                       */
/*                                      */
/****************************************/

int DecodeUnits( char units[],
                 enum unitTypes *baseUnits,
                 real *multiplier)
{
char cc, name[10];
int i;

*multiplier = 1.0;
for( i=0; i<2; i++)
    {
    strcpy( name, &(units[i]));
    *baseUnits = undefUnits;
    if( !strcmp( name, "ohm")) *baseUnits = ohms;
    if( !strcmp( name, "ohms")) *baseUnits = ohms;
    if( !strcmp( name, "mho")) *baseUnits = mhos;
    if( !strcmp( name, "mhos")) *baseUnits = mhos;
    if( !strcmp( name, "S")) *baseUnits = mhos;
    if( !strcmp( name, "a")) *baseUnits = amps;
    if( !strcmp( name, "A")) *baseUnits = amps;
    if( !strcmp( name, "f")) *baseUnits = farads;
    if( !strcmp( name, "F")) *baseUnits = farads;
    if( !strcmp( name, "h")) *baseUnits = henry;
    if( !strcmp( name, "H")) *baseUnits = henry;
    if( !strcmp( name, "v")) *baseUnits = volts;
    if( !strcmp( name, "V")) *baseUnits = volts;

    if( *baseUnits != undefUnits ) return(0);
    if( *multiplier != 1.0)
        {
```

```
        *multiplier = 0.0;
        return( -1);
        }
    cc = units[0];
    *multiplier = 0.0;
    if( cc == 'G') *multiplier = 1.0e9;
    if( cc == 'M') *multiplier = 1.0e6;
    if( (cc == 'K') || (cc == 'k') ) *multiplier = 1.0e3;
    if( cc == 'm') *multiplier = 1.0e-3;
    if( cc == 'u') *multiplier = 1.0e-6;
    if( cc == 'n') *multiplier = 1.0e-9;
    if( cc == 'p') *multiplier = 1.0e-12;
    if( *multiplier == 0.0 ) return( -1);
    }
return( 0);
}
/***********************************************/
/*                                             */
/*   DecodeOnePort()                           */
/*                                             */
/***********************************************/

int DecodeOnePort(    char token[][10],
                      enum unitTypes expectedUnits,
                      real values[],
                      char labels[][10],
                      int *jNode,
                      int *kNode)
{
real multiplier, rawValue;
enum unitTypes baseUnits;

DecodeUnits( &(token[3][0]), &baseUnits, &multiplier);
if( baseUnits != expectedUnits ) return( -20);
strcpy(labels[0],token[1]);
sscanf( &(token[2][0]), "%lg", &rawValue);
values[0] = multiplier * rawValue;
sscanf( &(token[4][0]),"%d", jNode);
sscanf( &(token[5][0]),"%d", kNode);
return( 0);
}
```

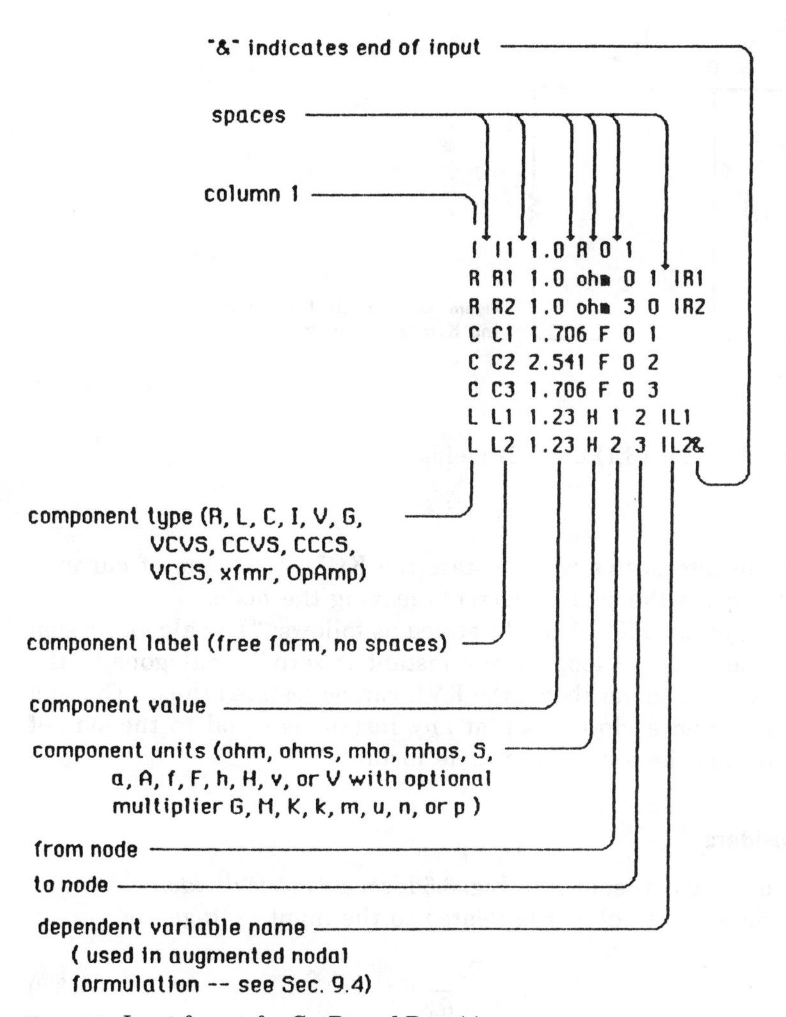

Figure 3.4 Input format for **GetBranchData( )**.

3.2 Kirchhoff's Law

Kirchhoff's current law (KCL) can be stated as follows: "The algebraic sum of currents into a node at any instant is zero." The word *algebraic* means that in the summation, currents leaving the node are treated as negative currents entering the node. Consider the circuit shown in Fig. 3.5. The sum of currents entering node a is given by

$$i_1 + (-i_2) + (-i_3) = 0 \tag{3.1}$$

The sum of currents entering node b is given by

$$(-i_1) + i_2 + i_3 = 0 \tag{3.2}$$

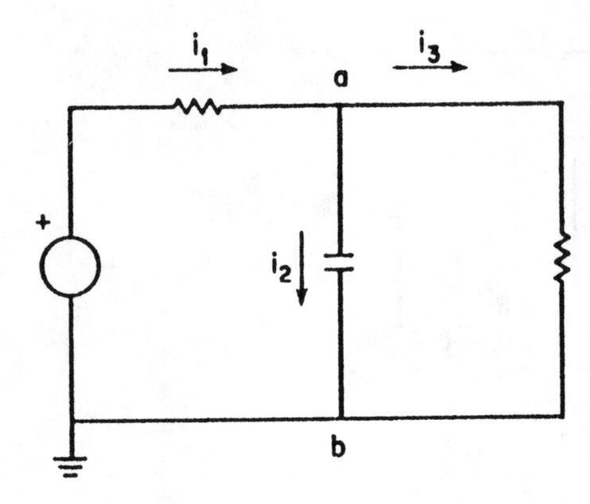

Figure 3.5 Circuit for illustrating Kirchhoff's current law.

Rearrangement of either (3.1) or (3.2) yields

$$i_1 = i_2 + i_3$$

which suggests an alternative way to state the KCL: "The sum of currents entering a node equals the sum of currents leaving the node."

Kirchhoff's voltage law (KVL) can be stated as follows: "The algebraic sum of voltages around a closed loop at any instant is zero." Analogous to the restatement of the KCL made above, the KVL can be restated thus: "The sum of voltage rises around a closed loop at any instant is equal to the sum of voltage drops around the loop at the same instant."

3.3 Voltage Dividers

Two resistors connected as shown in Fig. 3.6 form a *voltage divider*. When the current $I_2 = 0$, the output voltage is related to the input voltage via

$$V_{\text{out}} = \frac{R_2}{R_1 + R_2} V_{\text{in}} \tag{3.3}$$

Rarely, if ever, will I_2 be exactly zero; but in many practical applications I_2 will be very small compared to I_1 and thus conveniently approximated as zero. A more detailed analysis can be performed by using the model shown in Fig. 3.7. Analysis of the circuit shown reveals that

$$V_{\text{out}} = \frac{R_2}{R_1 + R_2 + R_1 R_2 / R_{\text{in}}} V_{\text{in}} \tag{3.4}$$

Note that as R_{in} becomes very large, Eq. (3.4) approaches Eq. (3.3).

3.4 Superposition and Reciprocity

Consider a network N which has an accessible input port and an observable loop, as shown in Fig. 3.8*a*. The applied voltage V_a produces a current of I_b

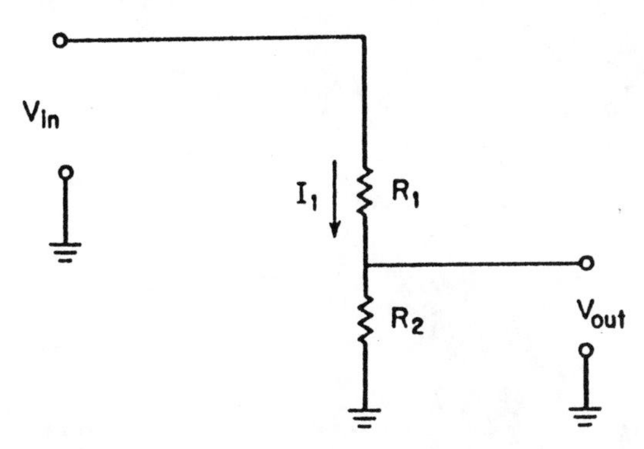

Figure 3.6 Voltage divider circuit.

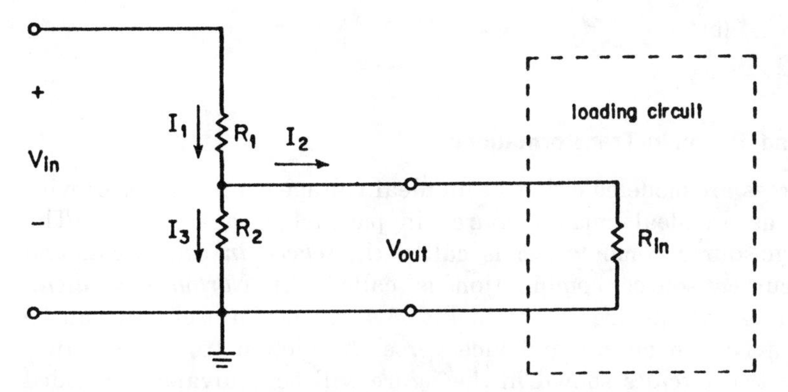

Figure 3.7 Circuit model for the analysis of a voltage divider.

in the observed loop. Now interchange the position of the source and the loop as shown in Fig. 3.8*b*. The applied voltage V_b produces a current of I_a in the observed loop. Provided that N contains only resistors, inductors, capacitors, and transformers, the ratio of V_a to I_b equals the ratio of V_b to I_a:

$$\frac{V_b}{I_a} = \frac{V_a}{I_b}$$

This is called the *principle of reciprocity*, and the network is said to be a *reciprocal* network.

Superposition

Consider a network which is simultaneously excited by a number of sources. The response to each individual source can be computed by replacing all other voltage sources with short circuits and all other current sources with open circuits. The response to all excitations applied simultaneously is equal to the sum of the individual responses thus computed.

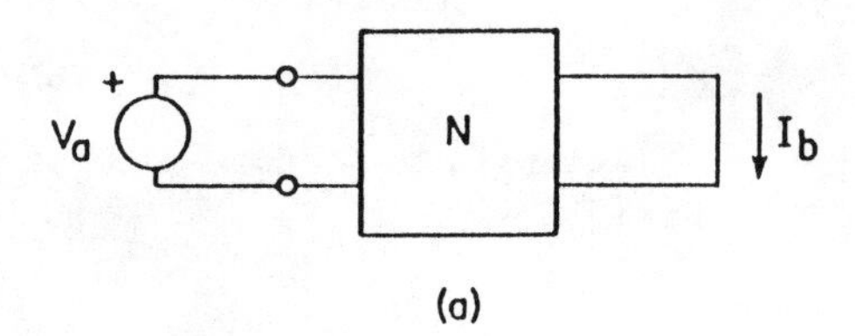

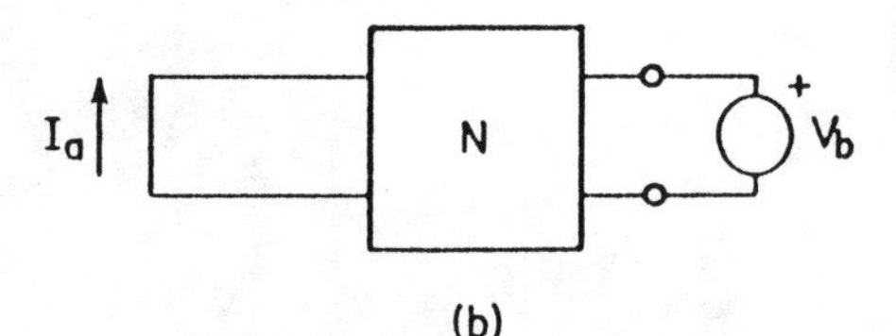

Figure 3.8 Illustration of reciprocity.

3.5 Norton and Thevenin Transformations

Practical sources are modeled either as an ideal voltage source in series with a resistor or as an ideal current source in parallel with a resistor. The resistor/voltage-source combination is called the *Thevenin equivalent*, and the resistor/current-source combination is called the *Norton equivalent*. These two configurations are shown in Fig. 3.9. A Thevenin circuit can be converted to a Norton circuit and vice versa. As viewed from the output terminals, the two circuits shown in the figure will be equivalent provided that $E = JR_s$.

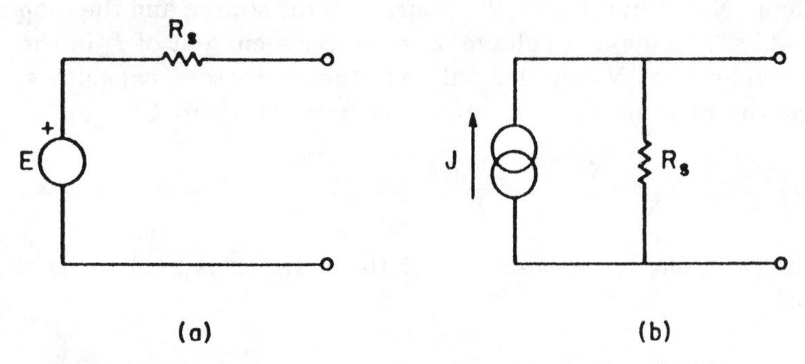

Figure 3.9 Models of practical sources: (*a*) Thevenin circuit and (*b*) Norton circuit.

Chapter

4

Solid-State Circuit Elements

This chapter is not intended as a tutorial on solid-state device physics or even solid-state circuit design. The primary motivation for this chapter is to show how semiconductor devices are modeled by using circuit elements that can be accommodated by various network analysis techniques. Some semiconductor device models include new types of ideal elements. Although semiconductor devices in general exhibit nonlinear behavior, some of the ideal elements used to model them are linear. These linear elements—controlled voltage sources and controlled current sources—are introduced in this chapter and included along with other linear ideal elements in the development of analysis techniques for linear networks appearing in subsequent chapters.

4.1 Additional Ideal Elements

Chapter 2 presented ideal resistors, capacitors, inductors, and transformers as well as independent voltage sources and independent current sources. This section presents a number of additional ideal elements that are needed to model the behavior of semiconductor devices.

Dependent sources

A dependent source is a voltage source or current source whose output is a function of a current or voltage measured elsewhere in the circuit. There are four possibilities:

1. Voltage-controlled current source (VCCS) or voltage-to-current transducer (VCT)
2. Voltage-controlled voltage source (VCVS) or voltage-to-voltage transducer (VVT)
3. Current-controlled current source (CCCS) or current-to-current transducer (CCT)

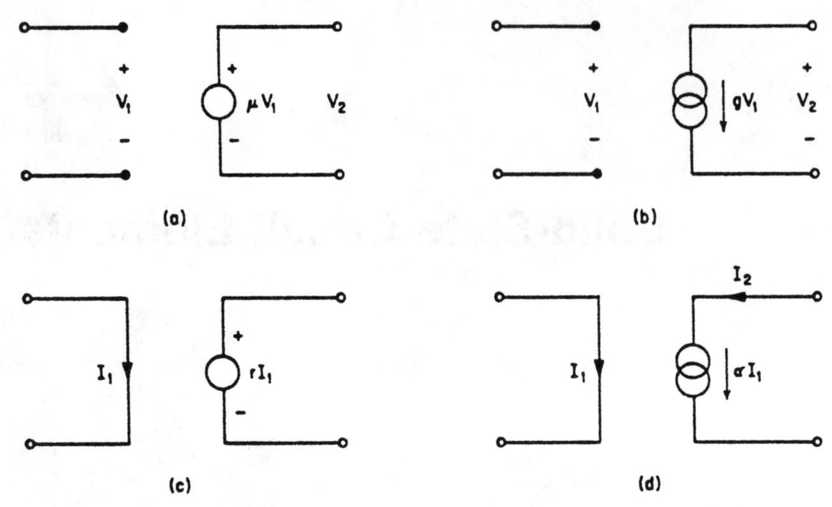

Figure 4.1 Dependent sources: (*a*) voltage-controlled voltage source (VCVS), (*b*) voltage-controlled current source (VCCS), (*c*) current-controlled voltage source (CCVS), (*d*) current-controlled current source (CCCS).

4. Current-controlled voltage source (CCVS) or current-to-voltage transducer (CVT)

The schematic symbols for dependent sources are shown in Fig. 4.1. Dependent sources can be either linear or nonlinear. In linear dependent sources, the controlled variable is related to the controlling variable via a simple multiplicative constant. Nonlinear dependent sources can have more complicated relationships between the controlling and controlled variables. [For example, see Eq. (4.1).]

4.2 Diodes

An ideal diode is a polarized device that exhibits zero resistance when *forward-biased* and infinite resistance when *reverse-biased*. The *V-I* characteristic for an ideal diode is only approximated by real-world semiconductor diodes. The *V-I* characteristic of a semiconductor diode containing an abrupt PN junction is given by

$$I = I_s\left(\exp\left(\frac{V}{V_T}\right) - 1\right) \tag{4.1}$$

where $V_T = kT/q$
q = charge on an electron = 1.6022×10^{-19} C
k = Boltzmann's constant = 1.3806×10^{-23} J/K
T = temperature of junction, K

The *saturation current* I_s is a device-dependent parameter that typically ranges between 10^{-9} and 10^{-6}. A typical *V-I* characteristic for a diode is sketched in

Fig. 4.2. Notice that for large negative values of V, the current is very nearly equal to $-I_s$. At room temperature (20°C), $V_T \approx 25.25\,\text{mV}$.

An equivalent circuit for modeling a diode junction is shown in Fig. 4.3. This circuit models only the semiconductor itself—elements must be added to model the effects of packaging, load inductance, etc., if these effects are significant. This model includes four parameters. These parameters in turn depend upon other parameters which must be either measured in the laboratory or obtained from manufacturers' data sheets.

R_b = semiconductor bulk resistance

R_s = junction leakage resistance

C_d = junction diffusion capacitance

$$= \frac{\tau q}{2mkT}(I + I_s)$$

m = curvature coefficient

τ = carrier lifetime

C_t = transition capacitance at $V = 0$

V_z = junction contact potential

N = junction grading constant

$$I = I_s\left(\exp\frac{V}{mV_T} - 1\right)$$

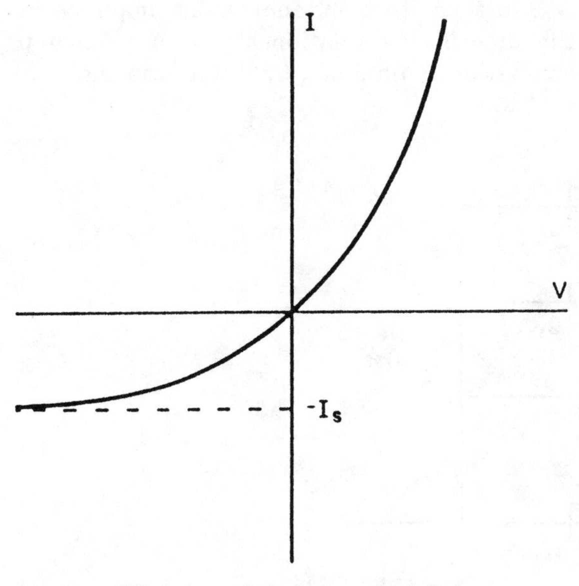

Figure 4.2 *V-I* characteristic for a typical diode.

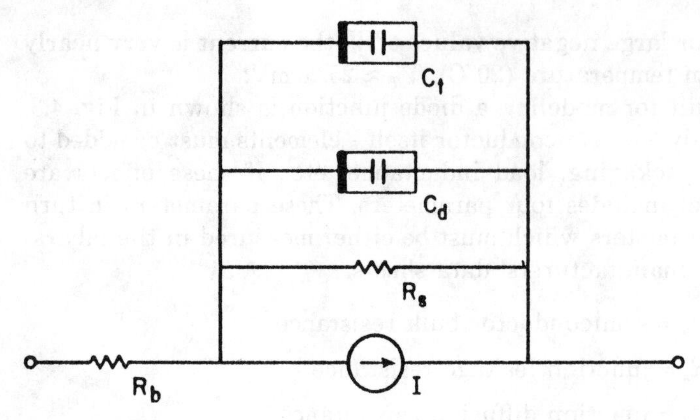

Figure 4.3 Circuit model of a junction diode.

The voltage-controlled current source in Fig. 4.3 can be replaced with a nonlinear resistor as shown in Fig. 4.4. The resistor has a *V-I* characteristic defined by

$$I_d = I_s\left(\exp\frac{V_d}{mV_T} - 1\right)$$

Often a diode in a circuit will be *biased* to a particular *operating point* on its *V-I* characteristic and then subjected to relatively small variations in voltage and current about this point. The operating-point current and voltage are denoted as I_0 and V_0. In this situation, analysis is simplified by using an approximation for the diode's *V-I* relationship at a particular operating point. Such an approximation is obtained by differentiating (4.1)

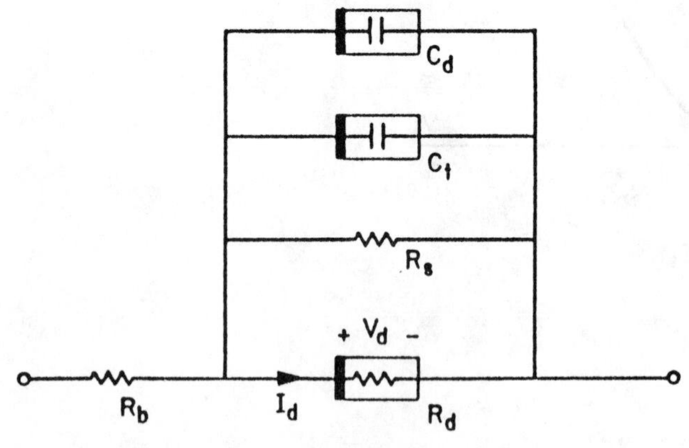

Figure 4.4 Alternative model of a junction diode.

and evaluating the result at the operating point

$$\left.\frac{dI}{dV}\right|_{V=V_0} = \frac{I_s}{V_0} \exp \frac{V_0}{V_T} \tag{4.2}$$

For any particular value of V_0, Eq. (4.2) yields a constant called the *dynamic conductance* of the diode at the operating point defined by V_0. This fact indicates that the small-signal behavior of the diode can be modeled as a conductance once the operating point is established.

4.3 Transistor Models

A *bipolar junction transistor* (BJT) can be connected in a *common-base*, *common-emitter*, or *common-collector* configuration, as shown in Fig. 4.5. The *hybrid* equivalent circuit or *h-parameter* equivalent circuit is shown in

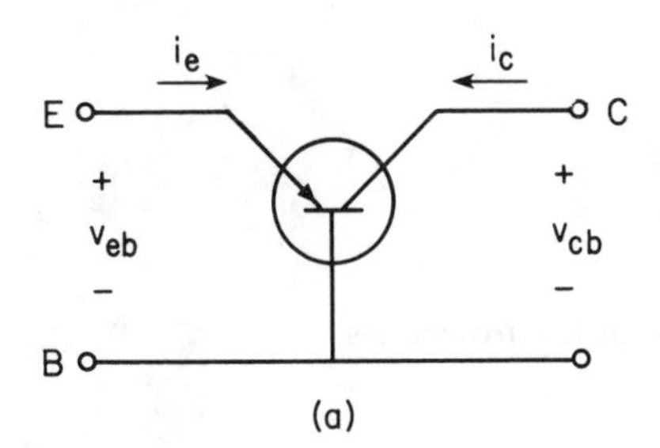

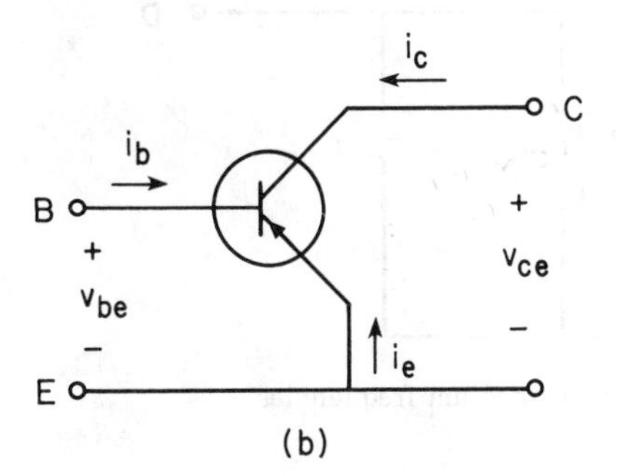

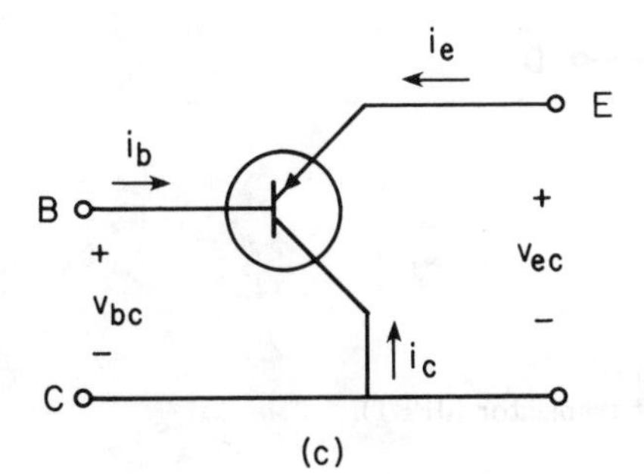

Figure 4.5 Three configurations for BJT circuits: (*a*) common-base, (*b*) common-emitter, (*c*) common-collector.

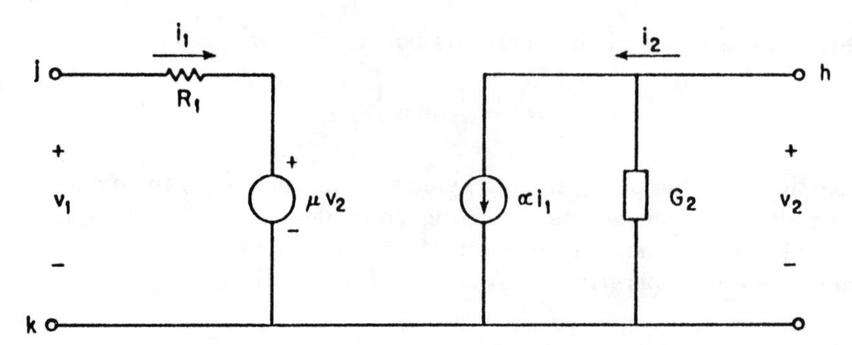

Figure 4.6 Hybrid equivalent circuit for a bipolar junction transistor.

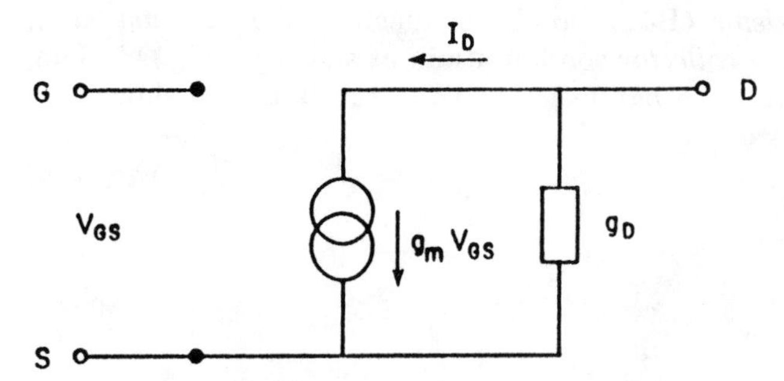

Figure 4.7 Small-signal model for a field-effect transistor at low frequencies.

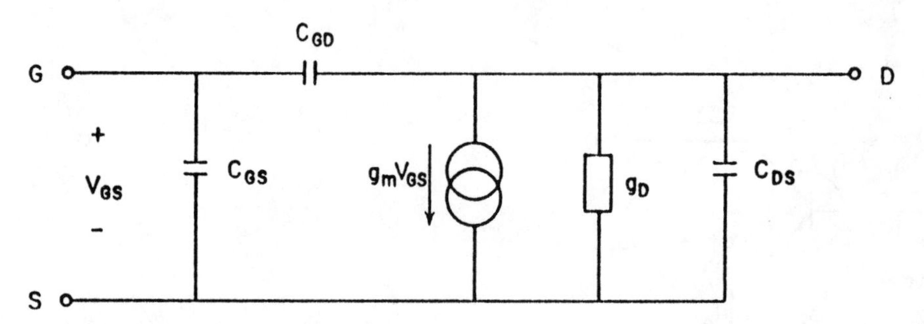

Figure 4.8 Small-signal model for a field-effect transistor at medium frequencies.

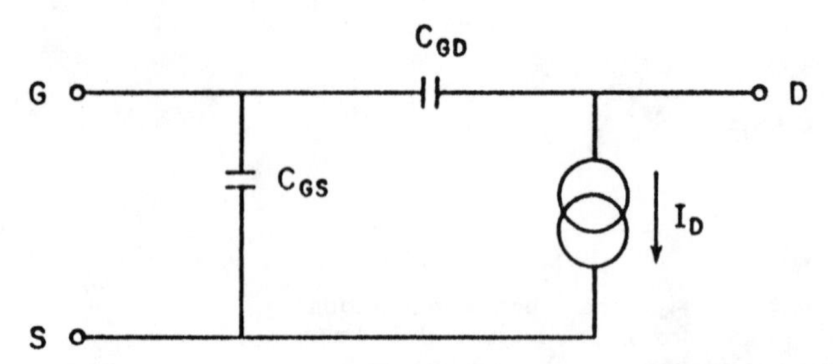

Figure 4.9 Large-signal model for a junction field-effect transistor (JFET).

TABLE 4.1 Parameter Assignments for Hybrid Equivalent Circuit

Generic parameter	Common base	Common emitter	Common collector
Terminal j	Emitter	Base	Base
Terminal k	Base	Emitter	Collector
Terminal n	Collector	Collector	Emitter
Input current i_1	i_e	i_b	i_b
Output current i_2	i_c	i_c	i_e
Input voltage v_1	v_{eb}	v_{be}	v_{bc}
Output voltage v_2	v_{cb}	v_{ce}	v_{ec}
Input impedance R_1	h_{ib}	h_{ie}	h_{ic}
Output admittance G_2	h_{ob}	h_{oe}	h_{oc}
Reverse voltage transfer μ	h_{rb}	h_{re}	h_{rc}
Forward current transfer α	h_{fb}	h_{fe}	h_{fc}

Fig. 4.6. The same equivalent-circuit topology can be used to model any of the three BJT configurations. The parameter assignments for each configuration are listed in Table 4.1.

Field-effect transistors

Three different circuits for modeling a *field-effect transistor* (FET) are shown in Figs. 4.7, 4.8, and 4.9.

Chapter

5

Circuit Specifications

This chapter presents the various ways in which circuits can be specified. The remainder of the book is devoted to computational methods for obtaining performance specifications of given circuits and for obtaining circuits which realize given specifications.

5.1 Networks as Linear Systems

In general, a *system* can be regarded as anything which accepts one or more input signals and operates upon them to produce one or more output signals. Many results from linear system theory are applicable to the study of circuits, since many circuits are in fact linear systems. When signals are represented as mathematical functions, it is convenient to represent systems as *operators* which operate upon input functions to produce output functions. Two alternative notations for representing a system H with input x and output y are given in Eqs. (5.1) and (5.2).

$$y = H[x] \tag{5.1}$$

$$y = Hx \tag{5.2}$$

This book will use the notation of Eq. (5.1) since this is less likely to be confused with multiplication of x by a value H.

In different presentations of system theory, the notational schemes used exhibit some variation. In this book, whenever it is necessary to emphasize the time-dependent nature of a system's input and output, Eq. (5.1) is written as

$$y(t) = H[x(t)]$$

Linearity

If the relaxed system H is *homogeneous*, multiplying the input by a constant gain is equivalent to multiplying the output by the same constant gain. Mathematically stated, the relaxed system H is homogeneous if, for constant a,

$$H[ax] = aH[x]$$

If the relaxed system H is *additive*, the output produced for the sum of two input signals is equal to the sum of the outputs produced for each input individually. Mathematically stated, the relaxed system H is additive if

$$H[x_1 + x_2] = H[x_1] + H[x_2]$$

A system that is both homogeneous and additive is said to "exhibit *superposition*" or to "satisfy the principle of superposition." A system which exhibits superposition is called a *linear* system. Under certain restrictions additivity implies homogeneity. Specifically the fact that a system H is additive implies that

$$H[\alpha x] = \alpha H[x]$$

for any rational α. Any real number can be approximated with arbitrary precision by a rational number; therefore, additivity implies homogeneity for real a provided that

$$\lim_{\alpha \to a} H[\alpha x] = H[ax]$$

Time invariance

A *time-invariant* system is a system whose characteristics do not change over time. A system is said to be *relaxed* if it is not still responding to any previously applied input. Given a relaxed system H such that

$$y(t) = H[x(t)]$$

then H is time-invariant if and only if

$$y(t - \tau) = H[x(t - \tau)]$$

for any τ and any $x(t)$. A time-invariant system is also called a *fixed*, or *stationary*, system. A system which is not time-invariant is called a *time-varying* system, *variable* system, or *nonstationary* system.

Causality

In a *causal* system, the output at time t can depend only upon the input at time t and prior. Mathematically stated, a system H is causal if and only if

$$H[x_1(t)] = H[x_2(t)] \qquad \text{for } t \le t_0$$

given that

$$x_1(t) = x_2(t) \qquad \text{for } t \le t_0$$

Put simply, a causal system "doesn't laugh until it's tickled." A *noncausal*, or *anticipatory*, system is one in which the present output does depend upon future values of the input. Noncausal systems occur in theory, but they cannot exist in the real world.

5.2 Characterization of Linear Systems

A linear system can be characterized by a differential equation, step response, impulse response, complex-frequency domain system function, or a transfer function. The relationships between these various characterizations are given in Table 5.1.

TABLE 5.1 Relationships between Characterizations of Linear Systems

Starting with	Perform	To obtain
Time-domain differential equation relating $x(t)$ and $y(t)$	Laplace transform	Complex-frequency domain system function
	Compute $y(t)$ for $x(t)$ = unit impulse	Impulse response $h(t)$
	Compute $y(t)$ for $x(t)$ = unit step	Step response $a(t)$
Step response $a(t)$	Differentiate with respect to time	Impulse response $h(t)$
Impulse response $h(t)$	Integrate with respect to time	Step response $a(t)$
	Laplace transform	Transfer function $H(s)$
Complex-frequency domain system function	Solve for $H(s) = Y(s)/X(s)$	Transfer function $H(s)$
Transfer function $H(s)$	Inverse Laplace transform	Impulse response $h(t)$

Impulse response

The *impulse response* of a system is the output response produced when a unit impulse $\delta(t)$ is applied to the input of the previously relaxed system. This is an especially convenient characterization of a linear system, since the response $y(t)$ to any continuous-time input signal $x(t)$ is given by

$$y(t) = \int_{-\infty}^{\infty} x(\tau)h(t, \tau)\,d\tau \tag{5.3}$$

where $h(t, \tau)$ denotes the system's response at time t to an impulse applied at time τ. The integral in (5.3) is sometimes referred to as the *superposition integral.* The particular notation used indicates that, in general, the system is time-varying. For a time-invariant system, the impulse response at time t depends only upon the time delay from τ to t; and we can redefine the impulse response to be a function of a single variable and denote it as $h(t - \tau)$. Equation (5.3) then becomes

$$y(t) = \int_{-\infty}^{\infty} x(\tau)h(t - \tau)\,d\tau \tag{5.4}$$

Via the simple change of variables $\lambda = t - \tau$, Eq. (5.4) can be rewritten as

$$y(t) = \int_{-\infty}^{\infty} x(t - \lambda)h(\lambda)\,d\lambda \tag{5.5}$$

If we assume that the input is zero for $t < 0$, the lower limit of integration can be changed to zero; and if we further assume that the system is causal, the upper limit of integration can be changed to t, thus yielding

$$y(t) = \int_{0}^{t} x(\tau)h(t - \tau)\,d\tau = \int_{0}^{t} x(t - \lambda)h(\lambda)\,d\lambda \tag{5.6}$$

The integrals in (5.6) are known as *convolution integrals*, and the equation indicates that $y(t)$ equals the *convolution* of $x(t)$ and $h(t)$. It is often more compact and convenient to denote this relationship as

$$y(t) = x(t) \otimes h(t) = h(t) \otimes x(t) \tag{5.7}$$

Various texts use different symbols, such as stars or asterisks, in place of "$\otimes$" to indicate convolution. The asterisk is probably favored by most printers, but in some contexts its use to indicate convolution could be confused with the complex conjugation operator. A typical system's impulse response is sketched in Fig. 5.1.

Step response

The *step response* of a system is the output signal produced when a unit step $u(t)$ is applied to the input of the previously relaxed system. Since the unit

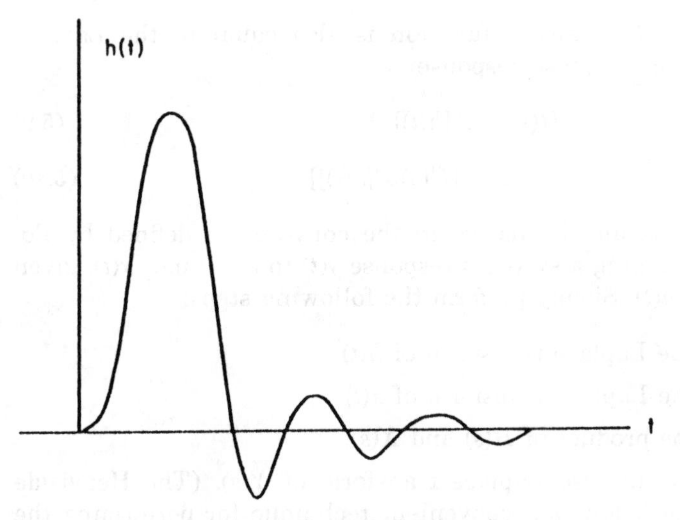

Figure 5.1 Impulse response of a typical system.

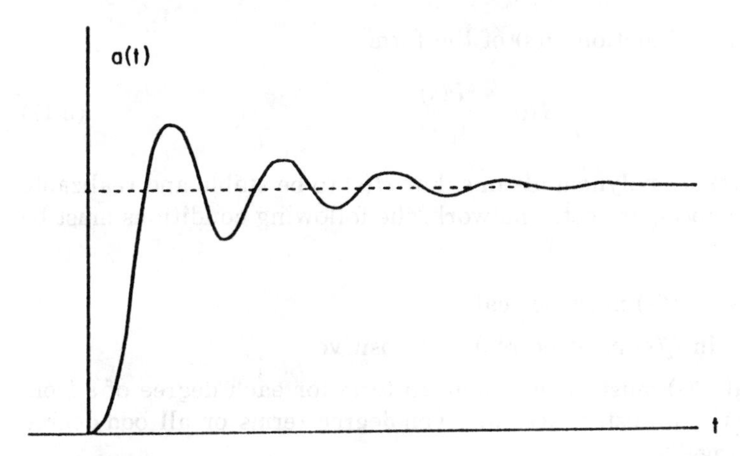

Figure 5.2 Step response of a typical system.

step is simply the time integration of a unit impulse, it can easily be shown that the step response of a system can be obtained by integrating the impulse response. A typical system's step response is shown in Fig. 5.2.

5.3 Transfer Functions

The *transfer function* $H(s)$ of a system is equal to the Laplace transform of the output signal divided by the Laplace transform of the input signal

$$H(s) = \frac{Y(s)}{X(s)} = \frac{\mathscr{L}[y(t)]}{\mathscr{L}[x(t)]} \tag{5.8}$$

It can be shown that the transfer function is also equal to the Laplace transform of the system's impulse response

$$H(s) = \mathscr{L}[h(t)] \tag{5.9}$$

Therefore,

$$y(t) = \mathscr{L}^{-1}[H(s)\mathscr{L}[x(t)]] \tag{5.10}$$

Equation (5.10) presents an alternative to the convolution defined by Eq. (5.6) of Sec. 5.2 for obtaining a system's response $y(t)$ to any input $x(t)$, given the impulse response $h(t)$. Simply perform the following steps:

1. Compute $H(s)$ as the Laplace transform of $h(t)$.
2. Compute $X(s)$ as the Laplace transform of $x(t)$.
3. Compute $Y(s)$ as the product of $H(s)$ and $X(s)$.
4. Compute $y(t)$ as the inverse Laplace transform of $Y(s)$. (The Heaviside expansion presented below is a convenient technique for performing the inverse transform operation.)

Consider a transfer function $H(s)$ of the form

$$H(s) = \frac{P(s)}{Q(s)} \tag{5.11}$$

where $P(s)$ and $Q(s)$ are polynomials in s. For $H(s)$ to be stable and realizable in the form of a lumped-parameter network, the following conditions must be satisfied:

1. The coefficients in $P(s)$ must be real.
2. The coefficients in $Q(s)$ must be real and positive.
3. The polynomial $Q(s)$ must have a nonzero term for each degree of s from the highest to the lowest, unless all even-degree terms or all odd-degree terms are missing.
4. If $H(s)$ is the voltage ratio or current ratio (i.e., the input and output are either both voltages or both currents), the maximum degree of s in $P(s)$ cannot exceed the maximum degree of s in $Q(s)$.
5. If $H(s)$ is a transfer impedance (i.e., the input is a current and the output is a voltage) or a transfer admittance (i.e., the input is a voltage and the output is a current), then the maximum degree of s in $P(s)$ can exceed the maximum degree of s in $Q(s)$ by at most 1.

Note that conditions 4 and 5 establish only upper limits on the degree of s in $P(s)$; in either case, the maximum degree of s in $P(s)$ may be as small as zero. Also note that these are necessary, but not sufficient, conditions for $H(s)$ to be a valid transfer function. A candidate $H(s)$ satisfying all these conditions may still not be realizable as a lumped-parameter network.

Example 5.1 Consider the following alleged transfer functions:

$$H_1(s) = \frac{s^2 - 2s + 1}{s^3 - 3s^2 + 3s + 1} \tag{5.12}$$

$$H_2(s) = \frac{s^4 + 2s^3 + 2s^2 - 3s + 1}{s^3 + 3s^2 + 3s + 2} \tag{5.13}$$

$$H_3(s) = \frac{s^2 - 2s + 1}{s^3 + 3s^2 + 1} \tag{5.14}$$

solution Equation (5.12) is not acceptable because the coefficient of s^2 in the denominator is negative. If Eq. (5.13) is intended as a voltage or current transfer ratio, it is not acceptable because the degree of the numerator exceeds the degree of the denominator. However, if Eq. (5.13) represents a transfer impedance or transfer admittance, it may be valid since the degree of the numerator exceeds the degree of the denominator by just 1. Equation (5.14) is not acceptable because the term for s is missing from the denominator.

A system's transfer function can be manipulated to provide a number of useful characterizations of the system's behavior. These characterizations are listed in Table 5.2 and examined in more detail in subsequent sections.

Poles and zeros

As pointed out previously, the transfer function for a realizable linear time-invariant system can always be expressed as a ratio of polynomials in s:

$$H(s) = \frac{P(s)}{Q(s)} \tag{5.15}$$

TABLE 5.2 System Characterizations Obtained from the Transfer Function

Starting with	Perform	To obtain
Transfer function $H(s)$	Compute roots of $H(s)$ denominator	Pole locations
	Compute roots of $H(s)$ numerator	Zero locations
	Compute $\lvert H(j\omega)\rvert$ over all ω	Magnitude response $A(\omega)$
	Compute $\arg\{H(j\omega)\}$ over all ω	Phase response $\theta(\omega)$
Phase response $\theta(\omega)$	Divide by ω	Phase delay $\tau_p(\omega)$
	Differentiate with respect to ω	Group delay $\tau_g(\omega)$

The numerator and denominator can each be factored to yield

$$H(s) = H_0 \frac{(s - z_1)(s - z_2)(s - z_3) \cdots (s - z_m)}{(s - p_1)(s - p_2)(s - p_3) \cdots (s - p_n)} \tag{5.16}$$

where the roots $z_1, z_2, \ldots, z_m$ of the numerator are called *zeros* of the transfer function and the roots $p_1, p_2, \ldots, p_n$ of the denominator are called *poles* of the transfer function. Together, poles and zeros can be collectively referred to as *critical frequencies.* Each factor $s - z_i$ is called a *zero factor,* and each factor $s - p_i$ is called a *pole factor.* A repeated zero appearing n times is called either an *nth-order zero* or a *zero of multiplicity n.* Likewise, a repeated pole appearing n times is called an *nth-order pole* or a *pole of multiplicity n.* Nonrepeated poles or zeros are sometimes described as *simple* or *distinct* to emphasize their nonrepeated nature.

Example 5.2 Consider the transfer function given by

$$H(s) = \frac{s^3 + 5s^2 + 8s + 4}{s^3 + 13s^2 + 59s + 87} \tag{5.17}$$

The numerator and denominator can be factored to yield

$$H(s) = \frac{(s + 2)^2(s + 1)}{(s + 5 + 2j)(s + 5 - 2j)(s + 3)} \tag{5.18}$$

Examination of (5.18) reveals that

$s = -1$ is a simple zero.

$s = -2$ is a second-order zero.

$s = -5 + 2j$ is a simple pole.

$s = -5 - 2j$ is a simple pole.

$s = -3$ is a simple pole.

A system's poles and zeros can be depicted graphically as locations in a complex plane, as shown in Fig. 5.3. In mathematics, the complex plane itself is called the *gaussian plane,* while a plot depicting complex values as points in the plane is called an *Argand diagram* or a *Wessel-Argand-Gauss diagram.* In the 1798 transactions of the Danish academy, Caspar Wessel (1745–1818) published a technique for graphical representation of complex numbers, and Jean Robert Argand published a similar technique in 1806. Geometric interpretation of complex numbers played a central role in the doctoral thesis of Gauss.

Pole locations can provide convenient indications of a system's behavior, as indicated in Table 5.3. Furthermore, poles and zeros possess the following properties, which can sometimes be used to expedite the analysis

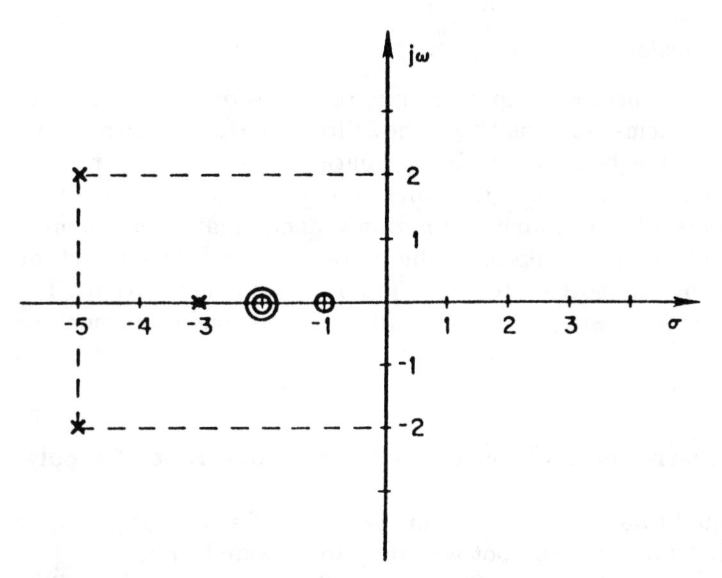

Figure 5.3 Plot of pole and zero locations.

TABLE 5.3 Impact of Pole Locations upon System Behavior

Pole type	Corresponding natural response component	Corresponding description of system behavior
Single real, negative	Decaying exponential	Stable
Single real, positive	Divergent exponential	Divergent instability
Real pair, negative, unequal	Decaying exponential	Overdamped (stable)
Real pair, negative, equal	Decaying exponential	Critically damped (stable)
Complex conjugate pair with negative real parts	Exponentially decaying sinusoid	Underdamped (stable)
Complex conjugate pair with zero real parts	Sinusoid	Undamped (marginally stable)
Complex conjugate pair with positive real parts	Exponentially saturating sinusoid	Oscillatory instability

of a system:

1. For real $H(s)$, complex or imaginary poles and zeros will each occur in complex conjugate pairs which are symmetric about the σ axis.
2. For $H(s)$ having even symmetry, the poles and zeros will exhibit symmetry about the $j\omega$ axis.
3. For nonnegative $H(s)$, any zeros on the $j\omega$ axis will occur in pairs.

5.4 Manipulating Transfer Functions

In many situations, it is necessary to determine the poles of a given transfer function. For some systems, such as Chebyshev filters or Butterworth filters, explicit expressions have been found for evaluation of pole locations. For other systems, such as Bessel filters, the poles must be found by numerically solving for the roots of the transfer function's denominator polynomial. Several root-finding algorithms appear in the literature, but I have found the *Laguerre method* to be the most useful for approximating pole locations. The approximate roots can be subjected to small-step iterative refinement or *polishing* as needed.

Algorithm 5.1. **[Laguerre method for approximating one root of a polynomial $P(z)$]**

Step 1. Set z equal to an initial guess for the value of a root. Typically, z is set to zero so that the smallest root will tend to be found first.

Step 2. Evaluate the polynomial $P(z)$ and its first two derivatives $P'(z)$ and $P''(z)$ at the current value of z.

Step 3. If $P(z)$ evaluates to zero or to within some predefined epsilon of zero, exit with the current value of z as the root. Otherwise, continue on to step 4.

Step 4. Compute a correction term Δz, using

$$\Delta z = \frac{N}{F \pm \sqrt{(N-1)(NG - G^2)}}$$

where $F \equiv P'(z)/P(z)$, $G \equiv F^2 - P''(z)/P(z)$, and the sign in the denominator is taken so as to minimize the magnitude of the correction (or, equivalently, so as to maximize the denominator).

Step 5. If the correction term Δz has a magnitude smaller than some specified fraction of the magnitude of z, then take z as the value of the root and terminate the algorithm.

Step 6. If the algorithm has been running for a while (let's say six iterations) and the correction value has gotten bigger since the previous iteration, then take z as the value of the root and terminate the algorithm.

Step 7. If the algorithm was not terminated in step 3, 5, or 6, then subtract Δz from z and go back to step 2. ■

A C routine **LaguerreMethod( )** that implements Algorithm 5.1 is provided in Listing 5.1.

```
/**************************************************/
/*                                                */
/*  Listing 5.1                                   */
/*                                                */
/*  LaguerreMethod()                              */
/*                                                */
/**************************************************/
#include "globDefs.h"
#include "protos.h"
extern FILE *fptr;
int LaguerreMethod(
                    int order,
                    struct complex coef[],
                    struct complex *zz,
                    real epsilon,
                    real epsilon2,
                    int maxIterations)
{
int iteration, j;
struct complex d2P_dz2, dP_dz, P, f, g, fSqrd, radical, cwork;
struct complex z, fPlusRad, fMinusRad, deltaZ;
real error, magZ, oldMagZ, fwork;
double dd1, dd2;

z = *zz;
oldMagZ = cAbs(z);

for( iteration=1; iteration<=maxIterations; iteration++)
    {
    d2P_dz2 = cmplx(0.0, 0.0);
    dP_dz = cmplx(0.0, 0.0);
    P = coef[order];
    error = cAbs(P);
    magZ = cAbs(z);
for( j=order-1; j>=0; j--)
    {
    d2P_dz2 = cAdd(dP_dz, cMult(z, d2P_dz2));
    dP_dz = cAdd( P, cMult(dP_dz,z));
    cwork = cMult(P,z);
    P = cAdd( coef[j], cMult(P,z));
    error = cAbs(P) + magZ * error;
    }
error = epsilon2 * error;
d2P_dz2 = sMult(2.0, d2P_dz2);
```

```
if( cAbs(P) < error)
    {
    *zz = z;
    return 1;
    }
f = cDiv( dP_dz,P);
fSqrd = cMult( f, f);
g = cSub( fSqrd, cDiv( d2P_dz2,P));
radical = cSub( sMult( (real)order, g), fSqrd);
fwork = (real)(order-1);
radical = cSqrt( sMult(fwork, radical));
fPlusRad = cAdd(f, radical);
fMinusRad = cSub( f, radical);
if( (cAbs(fPlusRad)) > (cAbs(fMinusRad)) )
    {
    deltaZ = cDiv( cmplx( (real)order, 0.0), fPlusRad);
    }
else
    {
    deltaZ = cDiv( cmplx( (real)order, 0.0), fMinusRad);
    }
z = cSub(z,deltaZ);
if( (iteration > 6)  && (cAbs(deltaZ) > oldMagZ) )
    {
    *zz = z;
    return 2;
    }
    if( cAbs(deltaZ) < ( epsilon * cAbs(z)))
        {
        *zz = z;
        return 3;
        }
    }
fprintf(fptr,"Laguerre method failed to converge \n");
return -1;
}
```

Derivatives of transfer functions

Algorithm 5.1 and the Heaviside expansion require evaluating derivatives of transfer functions. Consider the transfer function given by

$$H(s) = \frac{P(s)}{Q(s)} \qquad Q(s) \neq 0 \tag{5.19}$$

The derivative of $H(s)$ can be obtained as

$$\frac{d}{ds} H(s) = \frac{Q(s)\frac{d}{ds} P(s) - P(s)\frac{d}{ds} Q(s)}{[Q(s)]^2} \tag{5.20}$$

5.5 Partial Fraction Expansion

The *partial fraction expansion* of a transfer function is used in the Foster synthesis procedure presented in Sec. 10.2.

Algorithm 5.2. **(Partial fraction expansion)**

Step 1. Given a ratio of polynomials $N(s)/Q(s)$, compare the degree of N with the degree of Q. If the degree of N is less than the degree of Q, set $P(s) = N(s)$ and go to step 2. Otherwise, divide the numerator by the denominator until the degree of the remainder polynomial is less than the degree of Q. The result of this division will yield

$$\frac{N(s)}{Q(s)} = B_0 + B_1 s + B_2 s^2 + \cdots + B_{m-n} s^{m-n} + \frac{P(s)}{Q(s)}$$

where m = degree of $N(s)$
n = degree of $Q(s)$
$P(s)$ = remainder of division

Step 2. Find the factorization of $Q(s)$ such that

$$Q(s) = \prod_{k=1}^{n} (s - s_k) \tag{5.21}$$

where s_k, $k = 1, 2, \ldots, n$, are the n roots of $Q(s) = 0$. (The techniques of Sec. 5.4 may prove useful for this step.)

Step 3. If all the roots of $Q(s) = 0$ are simple, then the partial fraction expansion of $P(s)/Q(s)$ is

$$\frac{P(s)}{Q(s)} = \sum_{r=1}^{n} \frac{K_r}{s - s_r}$$

where

$$K_r = \left[\frac{(s - s_r)P(s)}{Q(s)}\right]_{s = s_r} \tag{5.22}$$

Step 4. If $Q(s) = 0$ has n distinct roots with each root s_r having a multiplicity of m_r, Eq. (5.21) can be written as

$$Q(s) = \prod_{k=1}^{n} (s - s_k)^{m_k} = (s - s_1)^{m_1}(s - s_2)^{m_2} \cdots (s - s_n)^{m_n} \tag{5.23}$$

and the partial fraction expansion is given by

$$\frac{P(s)}{Q(s)} = \sum_{r=1}^{n} \sum_{k=1}^{m_r} \frac{K_{rk}}{(s - s_r)^k}$$

where $$K_{rk} = \frac{1}{(k-1)!(m_r - k)!} \frac{d^{k-1}}{ds^{k-1}} \left[\frac{(s - s_r)m_r P(s)}{Q(s)} \right]_{s = s_r} \tag{5.24}$$

■

Heaviside expansion

Closely related to partial fraction expansion is the Heaviside expansion, which provides a straightforward computational method for obtaining the inverse Laplace transform of certain types of complex-frequency functions. The function to be operated on by the inverse transform must be expressed as a ratio of polynomials in s, where the order of the denominator polynomial exceeds the order of the numerator polynomial. If

$$H(s) = K_0 \frac{P(s)}{Q(s)} \tag{5.25}$$

where $Q(s)$ is in the form of Eq. (5.23), then inverse transformation via the Heaviside expansion yields

$$\mathcal{L}^{-1}[H(s)] = K_0 \sum_{r=1}^{n} \sum_{k=1}^{m_r} (K_{rk} t^{m_r - k} \exp s_r t) \tag{5.26}$$

where the K_{rk} are given by Eq. (5.24).

Simple pole case. The complexity of the expansion given in (5.26) is significantly reduced for the case of $Q(s)$ having no repeated roots. For this case, the denominator of (5.25) is given by

$$Q(s) = \prod_{k=1}^{n} (s - s_k) = (s - s_1)(s - s_2) \cdots (s - s_n) \qquad s_1 \neq s_2 \neq s_3 \neq \cdots \neq s_n \tag{5.27}$$

and inverse transformation via the Heaviside expansion yields

$$\mathcal{L}^{-1}[H(s)] = K_0 \sum_{r=1}^{n} K_r e^{s_r t} \tag{5.28}$$

where the K_r are given by Eq. (5.22).

The Heaviside expansion is named for Oliver Heaviside (1850–1925), an English physicist and electrical engineer who was the nephew of Charles Wheatstone (as in Wheatstone bridge).

5.6 Continued Fractions

A rational expression can be put into continued-fraction form by repeatedly performing a division followed by inversion of the remainder. Consider the following:

$$\frac{100}{63} = 1 + \frac{37}{63}$$

$$= 1 + \frac{1}{\frac{63}{37}}$$

$$= 1 + \frac{1}{1 + \frac{26}{37}}$$

$$= 1 + \frac{1}{1 + \frac{1}{\frac{37}{26}}}$$

$$= 1 + \frac{1}{1 + \frac{1}{1 + \frac{11}{26}}}$$

$$= 1 + \frac{1}{1 + \frac{1}{1 + \frac{1}{\frac{26}{11}}}}$$

$$= 1 + \frac{1}{1 + \frac{1}{1 + \frac{1}{2 + \frac{4}{11}}}}$$

$$= 1 + \frac{1}{1 + \frac{1}{1 + \frac{1}{2 + \frac{1}{2 + \frac{3}{4}}}}}$$

$$= 1 + \cfrac{1}{1 + \cfrac{1}{1 + \cfrac{1}{2 + \cfrac{1}{2 + \cfrac{1}{1 + \cfrac{1}{3}}}}}}$$

To ease the burden on typists and printers, the continued fraction developed above is often written as

$$\frac{100}{63} = 1 + \frac{1}{1+}\,\frac{1}{1+}\,\frac{1}{2+}\,\frac{1}{2+}\,\frac{1}{1+}\,\frac{1}{3}$$

The continued-fraction expansion of polynomial ratios corresponding to circuit functions is used extensively by the synthesis procedures presented in Chap. 10.

Manual procedure

Continued-fraction expansion of a polynomial ratio is performed by an alternating division-inversion process that is perhaps best introduced via a specific example. Consider the ratio of polynomials given by

$$X(s) = \frac{2s^3 - 5s^2 + s - 6}{3s^2 - s + 2}$$

To form the continued-fraction expansion of $X(s)$, we begin by performing the first step of the indicated division

$$\begin{array}{r|l} & \tfrac{2}{3}s \\ 3s^2 - s + 2 & 2s^3 - 5s^2 + s - 6 \\ & 2s^3 - \tfrac{2}{3}s^2 + \tfrac{4}{3}s \\ \hline & -\tfrac{13}{3}s^2 - \tfrac{1}{3}s - 6 \end{array}$$

The quotient and remainder thus produced can be used to represent $X(s)$ as

$$X(s) = \tfrac{2}{3}s + \frac{-\tfrac{13}{3}s^2 - \tfrac{1}{3}s - 6}{3s^2 - s + 2}$$

The second term can be inverted to yield

$$X(s) = \tfrac{2}{3}s + \frac{1}{(3s^2 - s + 2)/(-\tfrac{13}{3}s^2 - \tfrac{1}{3}s - 6)} \tag{5.29}$$

Next, we perform the first stage of the division indicated in the denominator of (5.29).

$$\begin{array}{r@{}l} & \quad -\tfrac{9}{13} \\ \frac{-13}{3}s^2 - \frac{1}{3}s - 6 \big) & 3s^2 - s \quad + 2 \\ & 3s^2 + \tfrac{3}{13}s + \tfrac{54}{13} \\ \hline & \quad - \tfrac{16}{13}s - \tfrac{28}{13} \end{array}$$

The ratio $X(s)$ can now be written as

$$X(s) = \frac{2}{3}s + \frac{1}{-\tfrac{9}{13} + (-\tfrac{16}{13}s - \tfrac{28}{13})/(-\tfrac{13}{3}s^2 - \tfrac{1}{3}s - 6)}$$

The second term in the denominator of the second term is then inverted to yield

$$X(s) = \frac{2}{3}s + \cfrac{1}{\cfrac{-9}{13} + \cfrac{1}{(-\tfrac{13}{3}s^2 - \tfrac{1}{3}s - 6)/(-\tfrac{16}{13}s - \tfrac{28}{13})}} \tag{5.30}$$

This is beginning to look like a continued fraction! We continue by performing the division indicated in the "bottommost" denominator and then inverting the "remainder-to-divisor" ratio in order to extend the continued fraction by one more level. This process continues until a division operation produces no remainder. The successive steps in the development of the continued-fraction representation for $X(s)$ are as follows:

$$X(s) = \frac{2}{3}s + \cfrac{1}{\cfrac{-9}{13} + \cfrac{1}{\cfrac{169}{48}s + \left(\cfrac{29}{4}s - 6\right)\Big/\left(-\cfrac{16}{13}s - \cfrac{28}{13}\right)}}$$

$$= \frac{2}{3}s + \cfrac{1}{\cfrac{-9}{13} + \cfrac{1}{\cfrac{169}{48}s + \cfrac{1}{\left(\cfrac{-16}{13}s - \cfrac{28}{13}\right)\Big/\left(\cfrac{29}{4}s - 6\right)}}}$$

$$= \frac{2}{3}s + \cfrac{1}{\cfrac{-9}{13} + \cfrac{1}{\cfrac{169}{48}s + \cfrac{1}{\cfrac{-64}{377} + \left(\cfrac{-29}{92}\right)\Big/\left(\cfrac{29}{4}s - 6\right)}}}$$

$$= \frac{2}{3}s + \cfrac{1}{\cfrac{-9}{13} + \cfrac{1}{\cfrac{169}{48}s + \cfrac{1}{\cfrac{-64}{377} + \cfrac{1}{-23s + (-6)\Big/\left(\cfrac{-29}{92}\right)}}}}$$

$$= \frac{2}{3}s + \cfrac{1}{\cfrac{-9}{13} + \cfrac{1}{\cfrac{169}{48}s + \cfrac{1}{\cfrac{-64}{377} + \cfrac{1}{-23s + \cfrac{1}{\left(\cfrac{-29}{92}\right)\Big/(-6)}}}}}$$

$$= \frac{2}{3}s + \cfrac{1}{\cfrac{-9}{13} + \cfrac{1}{\cfrac{169}{48}s + \cfrac{1}{\cfrac{-64}{377} + \cfrac{1}{-23s + \cfrac{1}{\cfrac{29}{552}}}}}}$$

This repetitive divide/invert procedure is fine for manual calculation of a continued-fraction expansion, but it is cumbersome and perhaps a bit too unstructured for efficient implementation as a computer program. Thus we are driven to develop a more general approach.

Computer approach

Consider the ratio of polynomials given by

$$Z(s) = \frac{A(s)}{B(s)}$$

where

$$A(s) = \sum_{k=0}^{n} a_k s^k \qquad B(s) = \sum_{k=0}^{m} b_k s^k$$

After the first step of the continued-fraction process (which consists of a single step of polynomial division), $Z(s)$ can be expressed in terms of a quotient $Q(s)$ and a remainder $R(s)$ as

$$Z(s) = Q(s) + \frac{R(s)}{B(s)} \tag{5.31}$$

where $Q(s) = \dfrac{a_n}{b_m} s^{n-m}$

$$R(s) = \sum_{k=0}^{n-1} c_k s^k$$

$$c_k = \frac{a_k b_m - a_n b_{m+k-n}}{b_m}$$

$$a_j, b_j \triangleq 0 \qquad \text{for } j < 0$$

The second step of the continued-fraction expansion entails inverting the second term of (5.31) and performing another single step of the indicated division, to yield

$$\begin{aligned} Z(s) &= Q(s) + \frac{1}{B(s)/R(s)} \\ \frac{B(s)}{R(s)} &= Q_2(s) + \frac{R_2(s)}{R(s)} \end{aligned} \tag{5.32}$$

where $Q_2(s) = \dfrac{b_m}{c_{n-1}} s^{m+1-n}$

$$R_2(s) = \sum_{k=0}^{m-1} d_k s^k$$

$$d_k = \frac{b_k c_{n-1} - b_m c_{n+k-m-1}}{c_{n-1}}$$

$$c_j \triangleq 0 \qquad \text{for } j < 0$$

The invert/divide process is repeated until a division produces a remainder of zero. Let the iteration represented by Eq. (5.31) be counted as iteration 1 and the iteration represented by Eq. (5.32) be counted as iteration 2. If subsequent iterations are numbered consecutively, it turns out that for iteration j, the remainder polynomial $R_j(s)$ is given by

$$R_j(s) = \sum_{k=0}^{h} r_{i,k} s^k \tag{5.33}$$

where

$$h = \begin{cases} n - \dfrac{j+1}{2} & j \text{ odd} \\[2ex] m - \dfrac{j}{2} & j \text{ even} \end{cases}$$

$$r_{[j][k]} = \begin{cases} \dfrac{r_{[j-2][k]} r_{[j-1][m-(j-1)/2]} - r_{[j-2][n-(j-1)/2]} r_{[j-1][m+k-n]}}{r_{[j-1][m-(j-1)/2]}} & j \text{ odd} \\[3ex] \dfrac{r_{[j-2][k]} r_{[j-1][n-j/2]} - r_{[j-2][m+1-j/2]} r_{[j-1][k+n-m-1]}}{r_{[j-1][n-j/2]}} & j \text{ even} \end{cases}$$

The notation in Eq. (5.33) has been modified to include two subscripts for each coefficient. The first subscript indicates the iteration number, and the second subscript indicates the power of s with which the coefficient is associated.

In a similar vein, the quotient produced by the jth iteration is given by

$$Q_j(s) = \begin{cases} \dfrac{r_{[j-2][n-(j-1)/2]}}{r_{[j-1][m-(j-1)/2]}} s^{n-m} & j \text{ odd} \\ \dfrac{r_{[j-2][m+1-j/2]}}{r_{[j-1][n-j/2]}} s^{m+1-n} & j \text{ even} \end{cases} \tag{5.34}$$

Equations (5.33) and (5.34) can be easily incorporated into a computer program for automated evaluation of continued-fraction expansions.

5.7 Frequency Response

A system's *steady-state response* $H(j\omega)$ can be determined by evaluating the transfer function $H(s)$ at $s = j\omega$:

$$H(j\omega) = |H(j\omega)|e^{j\theta(\omega)} = H(s)\big|_{s=j\omega} \tag{5.35}$$

The *magnitude response* is simply the magnitude of $H(j\omega)$:

$$|H(j\omega)| = ([\text{Re}\{H(j\omega)\}]^2 + [\text{Im}\{H(j\omega)\}]^2)^{1/2} \tag{5.36}$$

It can be shown that

$$|H(j\omega)|^2 = H(s)H(-s)\big|_{s=j\omega} \tag{5.37}$$

If $H(s)$ is available in factored form as given by

$$H(s) = H_0 \frac{(s-z_1)(s-z_2)(s-z_3)\cdots(s-z_m)}{(s-p_1)(s-p_2)(s-p_3)\cdots(s-p_n)} \tag{5.38}$$

then the magnitude response can be obtained by replacing each factor with its absolute value evaluated at $s = j\omega$:

$$|H(j\omega)| = H_0 \frac{|j\omega - z_1| \cdot |j\omega - z_2| \cdot |j\omega - z_3| \cdots |j\omega - z_m|}{|j\omega - p_1| \cdot |j\omega - p_2| \cdot |j\omega - p_3| \cdots |j\omega - p_n|} \tag{5.39}$$

The *phase response* $\theta(\omega)$ is given by

$$\theta(\omega) = \tan^{-1} \frac{\text{Im}\{H(j\omega)\}}{\text{Re}\{H(j\omega)\}} \tag{5.40}$$

Phase delay

The *phase delay* $\tau_p(\omega)$ of a system is defined as

$$\tau_p(\omega) = \frac{-\theta(\omega)}{\omega} \tag{5.41}$$

where $\theta(\omega)$ is the phase response defined in Eq. (5.40). When evaluated at any specific frequency ω_1, Eq. (5.41) will yield the time delay experienced by a sinusoid of frequency ω passing through the system. Some authors define $\tau_p(\omega)$ without the minus sign shown on the right-hand side (RHS) of (5.41). As illustrated in Fig. 5.4, the phase delay at a frequency ω_1 is equal to the negative slope of a secant drawn from the origin to the phase response curve at ω_1.

Group delay

The *group delay* $\tau_g(\omega)$ of a system is defined as

$$\tau_g(\omega) = \frac{-d}{dt}\,\theta(\omega) \tag{5.42}$$

where $\theta(\omega)$ is the phase response defined in (5.40). In the case of a modulated carrier passing through the system, the modulation envelope will be delayed by an amount which is, in general, not equal to the delay $\tau_p(\omega)$ that is experienced by the carrier. If the system exhibits constant group delay over the entire bandwidth of the modulated signal, then the envelope will be delayed by an amount equal to τ_g. If the group delay is not constant over the entire bandwidth of the signal, the envelope will be distorted. As shown in

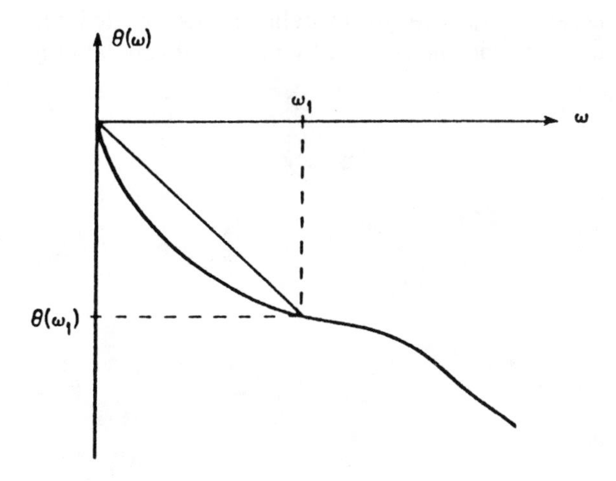

Figure 5.4 Phase delay.

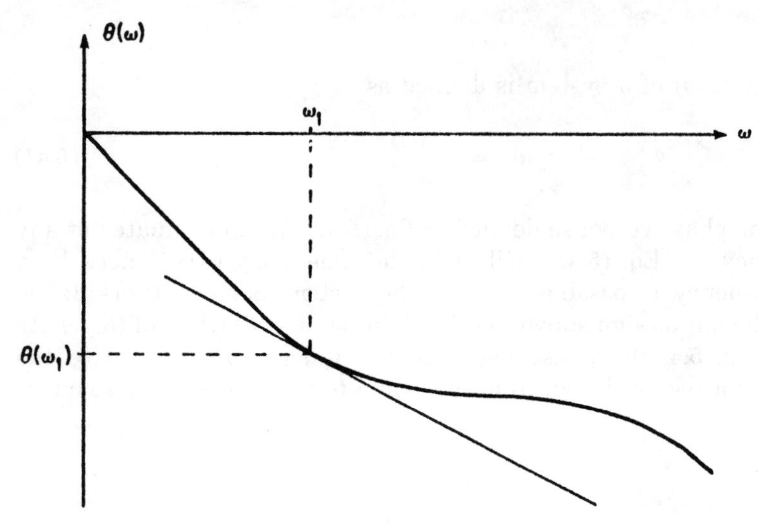

Figure 5.5 Group delay.

Fig. 5.5, the group delay at a frequency ω_1 is equal to the negative slope of a tangent to the phase response at ω_1.

Assuming that the phase response of a system is sufficiently smooth, it can be approximated as

$$\theta(\omega + \omega_c) = \tau_p \omega_c + \tau_g \omega_c \tag{5.43}$$

If an input signal $x(t) = a(t) \cos \omega_c t$ is applied to a system for which (5.43) holds, the output response will be given by

$$y(t) = Ka(t - \tau_g) \cos \omega_c (t - \tau_p) \tag{5.44}$$

Since the envelope $a(t)$ is delayed by τ_g, the group delay is also called the *envelope delay*. Likewise, since the carrier is delayed by τ_p, the phase delay is also called the *carrier delay*.

Chapter

6

Filter Specifications

Frequency-selective filtering represents perhaps the most common application of passive circuits. This chapter reviews the basics of filter specification and then examines in detail the characteristics of the three most widely used types of filters. The desired filter response is usually selected based upon requirements of the application, and then the corresponding transfer function is used to drive the circuit synthesis techniques presented in Chap. 10.

6.1 Filter Fundamentals

Ideal filters would have rectangular magnitude responses, as shown in Fig. 6.1. The desired frequencies are passed with no attenuation, while the undesired frequencies are completely blocked. If such filters could be implemented, they would enjoy widespread use. Unfortunately, ideal filters are noncausal and therefore not realizable. However, there are practical filter designs that approximate the ideal filter characteristics and which are realizable. Each of the major types—Butterworth, Chebyshev, and Bessel—optimizes a different aspect of the approximation.

Magnitude response features of lowpass filters

The magnitude response of a practical lowpass filter will usually have one of the four general shapes shown in Figs. 6.2 through 6.5. In all four cases the filter characteristics divide the spectrum into three general regions, as shown. The *passband* extends from direct current up to the cutoff frequency ω_c. The *transition band* extends from ω_c up to the beginning of the stopband at ω_1, and the *stopband* extends upward from ω_1 to infinity. The cutoff frequency ω_c is the frequency at which the amplitude response falls to a specified fraction [usually -3 decibels (dB), sometimes -1 dB] of the peak

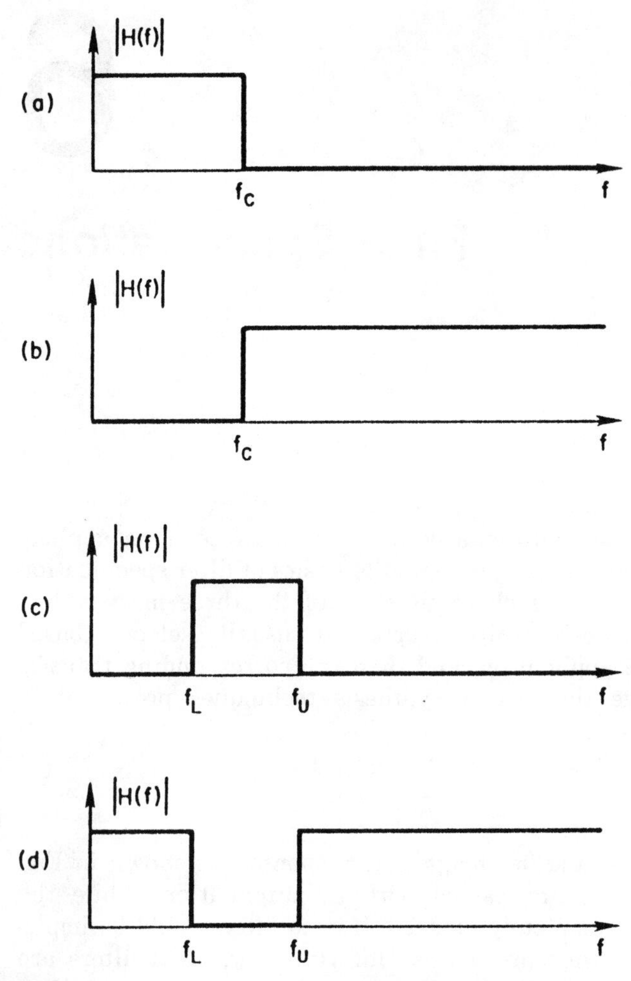

Figure 6.1 Ideal filter responses: (*a*) lowpass, (*b*) highpass, (*c*) bandpass, (*d*) bandstop.

passband values. Defining the frequency ω_1 which marks the beginning of the stopband is not quite so straightforward. In Fig. 6.2 or 6.3 there really isn't any particular feature that indicates just where ω_1 should be located. The usual approach involves specifying a *minimum stopband loss* α_2 (or conversely a maximum stopband amplitude A_2) and then defining ω_1 as the lowest frequency at which the loss exceeds and subsequently continues to exceed α_2. The width W_T of the transition band is equal to $\omega_c - \omega_1$. The quantity W_T/ω_c is sometimes called the *normalized transition width.* In the case of response shapes like those shown in Figs. 6.4 and 6.5, the minimum stopband loss is clearly defined by the peaks of the stopband ripples.

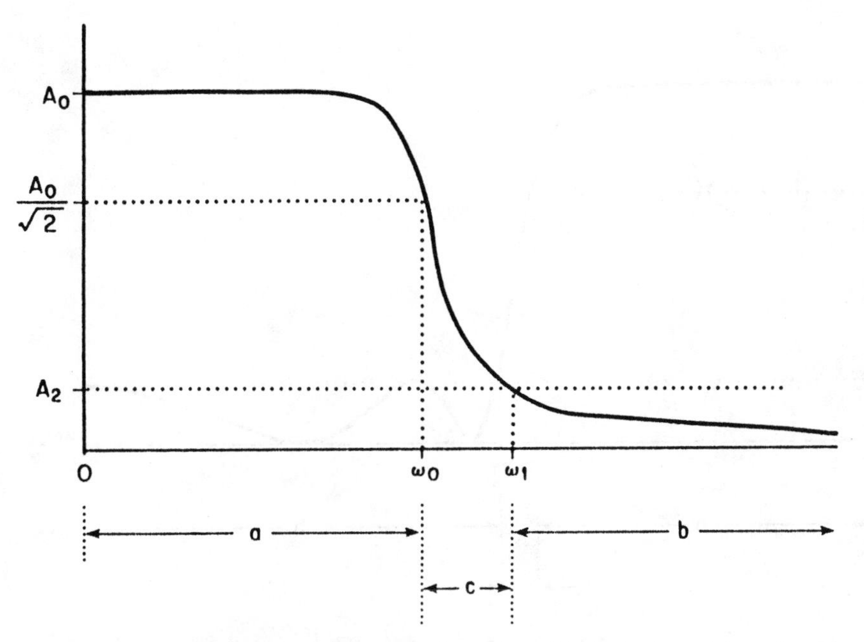

Figure 6.2 Monotonic magnitude response of a practical lowpass filter: (*a*) passband, (*b*) stopband, (*c*) transition band.

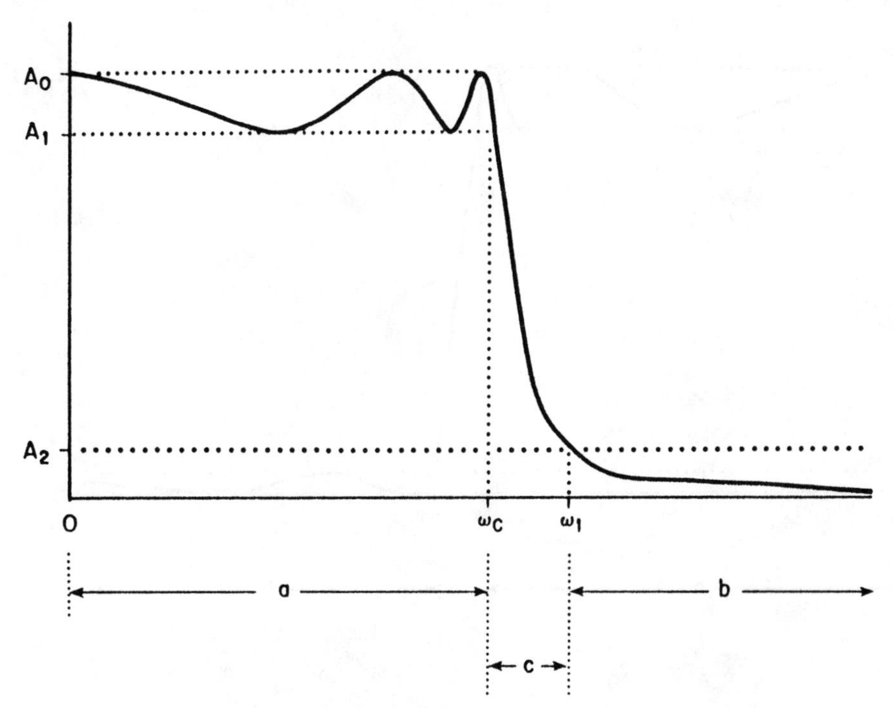

Figure 6.3 Magnitude response of a practical lowpass filter with ripples in the passband: (*a*) passband, (*b*) stopband, (*c*) transition band.

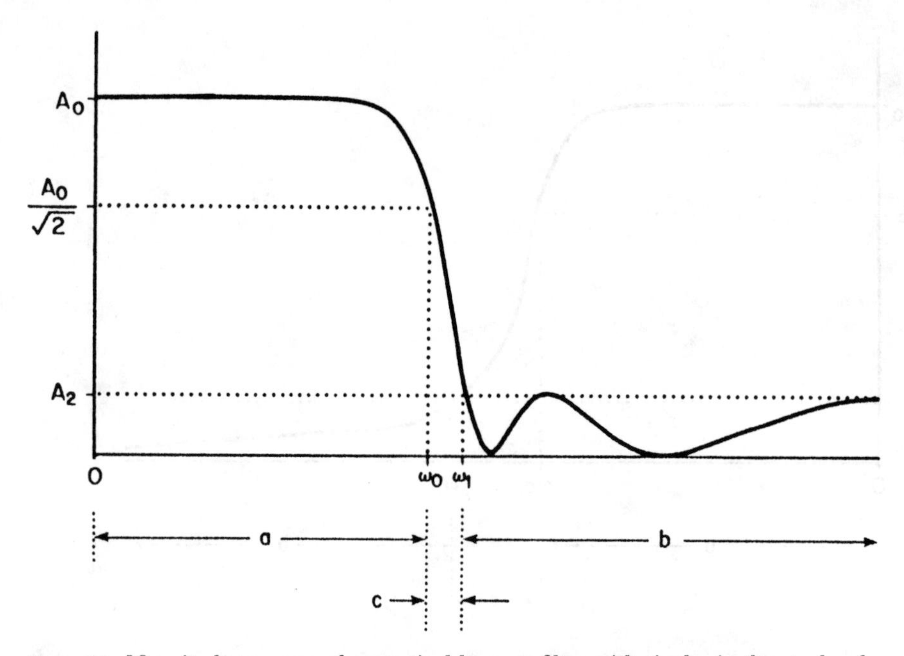

Figure 6.4 Magnitude response of a practical lowpass filter with ripples in the stopband: (*a*) passband, (*b*) stopband, (*c*) transition band.

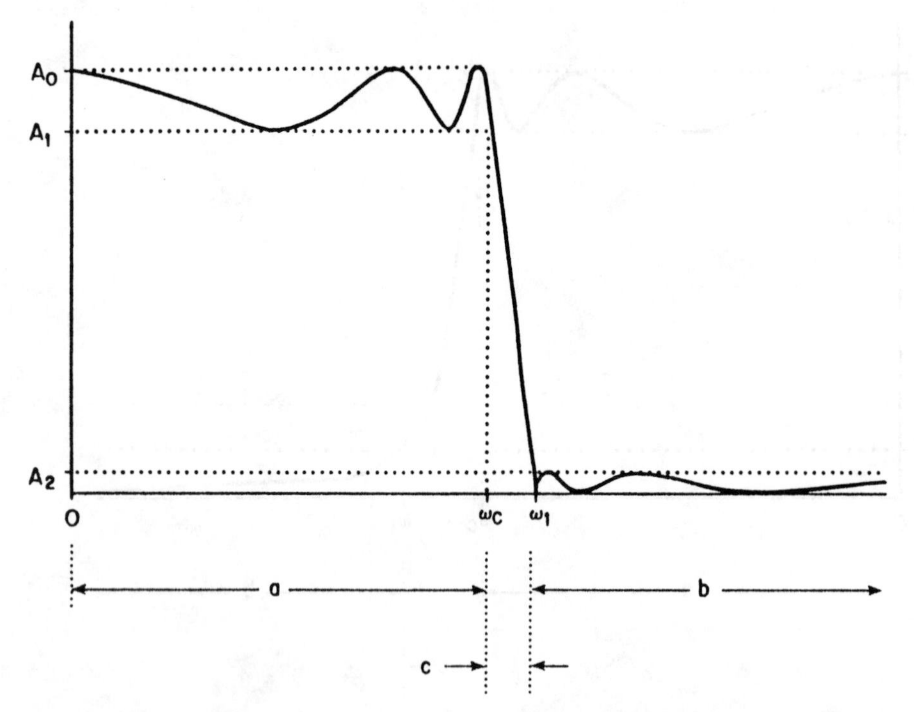

Figure 6.5 Magnitude response of a practical lowpass filter with ripples in the passband and stopband: (*a*) passband, (*b*) stopband, (*c*) transition band.

Scaling of lowpass filter responses

In plots of practical filter responses, the frequency axes are almost universally plotted on logarithmic scales. Magnitude response curves for lowpass filters are scaled so that the cutoff frequency occurs at a convenient frequency such as 1 radian per second (rad/s), 1 hertz (Hz), or 1 kilohertz (kHz). A single set of such normalized curves can then be denormalized to fit any particular cutoff requirement.

Magnitude scaling. The vertical axes of a filter's magnitude response can be presented in several different forms. In theoretical presentations, the magnitude response is often plotted on a linear scale. In practical design situations, it is convenient to work with plots of attenuation in decibels using a high-resolution linear scale in the passband and a lower-resolution linear scale in the stopband. This allows details of the passband response to be shown as well as large attenuation values deep into the stopband. In nearly all cases, the data is normalized to present a 0-dB attenuation at the peak of the passband.

Phase response. The phase response is plotted as a phase angle, in degrees or radians, versus frequency. By adding or subtracting the appropriate number of full-cycle offsets (that is, 2π rad or 360°), the phase response can be presented either as a single curve extending over several full cycles (Fig. 6.6) or as an equivalent set of curves, each of which extends over a single cycle (Fig. 6.7). Phase calculations will usually yield results confined to a single 2π cycle. Listing 6.1 contains a C function **UnwrapPhase( )** that can be used to convert such data to the multicycle form of Fig. 6.6.

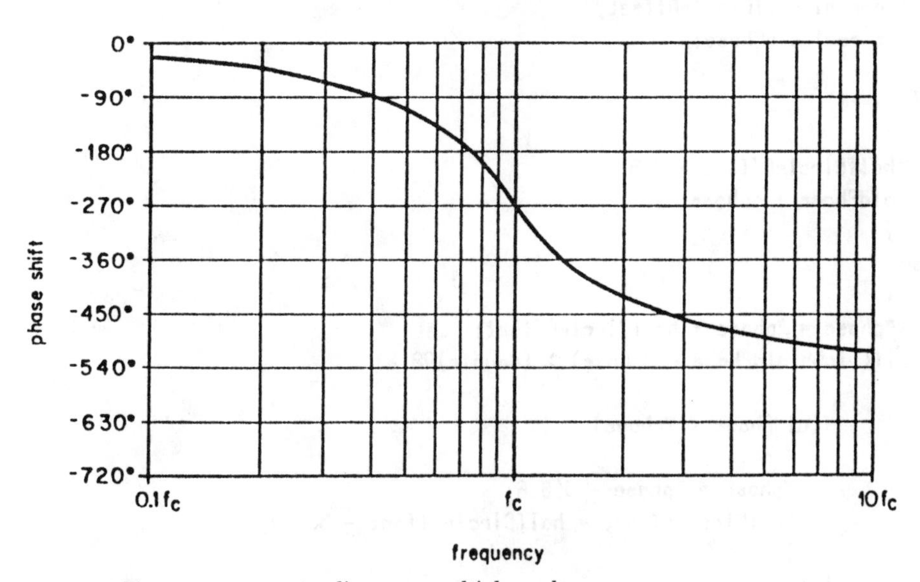

Figure 6.6 Phase response extending over multiple cycles.

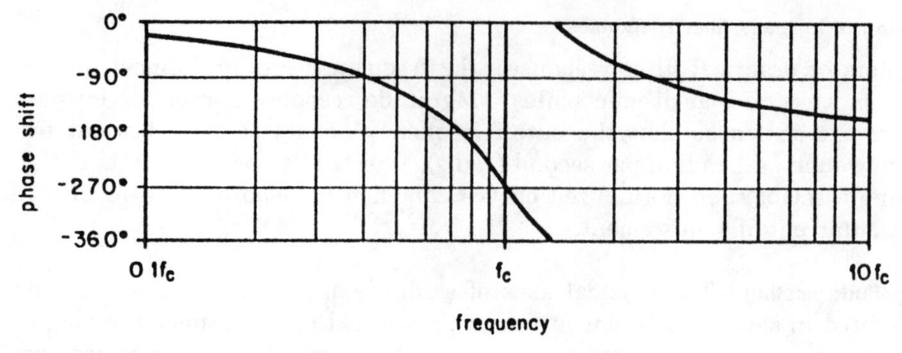

Figure 6.7 Phase response confined to a single-cycle range.

```
/***********************************/
/*                                 */
/*   Listing 6.1                   */
/*                                 */
/*   UnwrapPhase()                 */
/*                                 */
/***********************************/
#include <math.h>

void UnwrapPhase( int ix,
                  real *phase)
{
static real halfCircleOffset;
static real oldPhase;

if( ix==0)
    {
    halfCircleOffset = 0.0;
    oldPhase = *phase;
    }
else
    {
    *phase = *phase + halfCircleOffset;
    if( fabs(oldPhase - *phase) > (double)90.0)
        {
        if(oldPhase < *phase)
            {
            *phase = *phase - 360.0;
            halfCircleOffset = halfCircleOffset - 360.0;
            }
        else
```

```
                {
                *phase = *phase + 360.0;
                halfCircleOffset = halfCircleOffset + 360.0;
                }
            }
        oldPhase = *phase;
        }
    return;
    }
```

Step response. Normalized step response plots are obtained by computing the step response from the normalized transfer function. The inherent scaling of the time axis will thus depend upon the transient characteristics of the normalized filter. The amplitude axis scaling is not dependent upon normalization. The usual lowpass presentation will require that the response be denormalized by dividing the frequency axis by some form of the cutoff frequency.

Impulse response. Normalized impulse response plots are obtained by computing the impulse response from the normalized transfer function. Since an impulse response will always have an area of unity, both the time axis and the amplitude axis will exhibit inherent scaling which depends upon the transient characteristics of the normalized filter. The usual lowpass presentation will require that the response be denormalized by multiplying the amplitude by some form of the cutoff frequency and dividing the time axis by the same factor.

Highpass filters

Highpass filters are usually designed via transformation of lowpass designs. Normalized lowpass transfer functions can be converted to corresponding highpass transfer functions by simply replacing each occurrence of s with $1/s$. This will cause the magnitude response to be "flipped" around a line at f_c, as shown in Fig. 6.8. (Note that this "flip" works only when the frequency is plotted on a logarithmic scale.) Rather than actually trying to draw a flipped response curve, it is much simpler to take the reciprocals of all the important frequencies for the highpass filter in question and then read the appropriate response directly from the lowpass curves.

Bandpass filters

Bandpass filters are classified as wideband or narrowband based upon the relative width of their passbands. Different methods are used for obtaining the transfer function for each type.

Wideband bandpass filters. Wideband bandpass filters can be realized by cascading a lowpass filter and a highpass filter. This approach will be acceptable

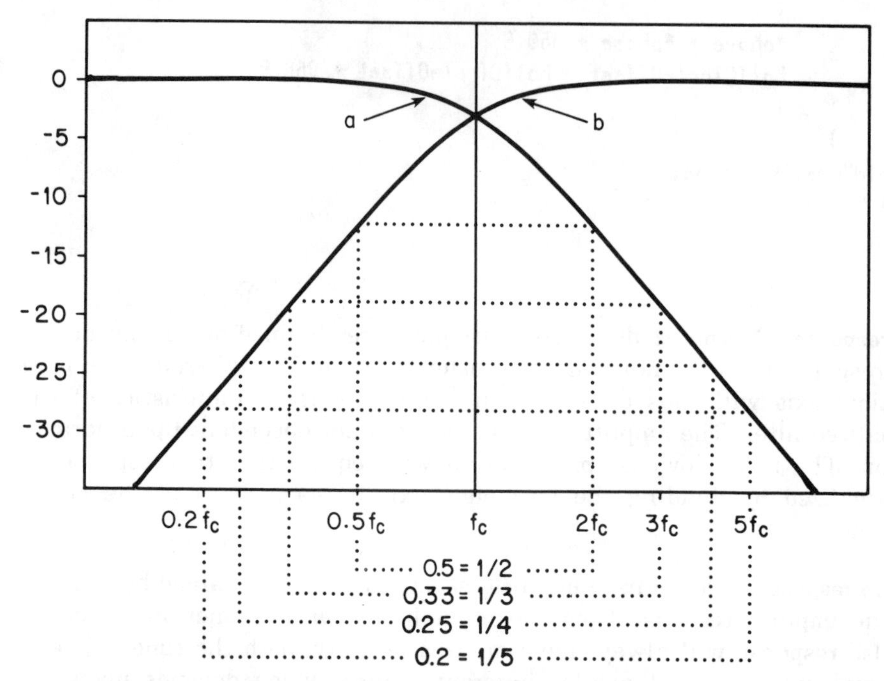

Figure 6.8 Relationship between lowpass and highpass magnitude responses: (*a*) lowpass response and (*b*) highpass response.

as long as the bandpass filters used exhibit relatively sharp transitions from the passband to cutoff. Relatively narrow bandwidths and/or gradual rolloffs which begin within the passband can cause a significant centerband loss, as shown in Fig. 6.9. In situations where such losses are unacceptable, other bandpass filter realizations must be used. A general rule of thumb is to use narrowband techniques for passbands which are an octave or smaller.

Narrowband bandpass filters. A normalized lowpass filter can be converted to a normalized narrowband bandpass filter by substituting $s - 1/s$ for s in the lowpass transfer function. The center frequency of the resulting bandpass filter will be at the cutoff frequency of the original lowpass filter, and the passband will be symmetric about the center frequency when it is plotted on a logarithmic frequency scale. At any particular attenuation level, the bandwidth of the bandpass filter will equal the frequency at which the lowpass filter exhibits the same attenuation. (See Fig. 6.10.) This particular bandpass transformation preserves the magnitude response shape of the lowpass prototype, but distorts the transient responses.

Bandstop filters. A normalized lowpass filter can be converted to a normalized bandstop filter by substituting $s/(s^2 - 1)$ for s in the lowpass transfer function. The center frequency of the resulting bandstop filter will be at the

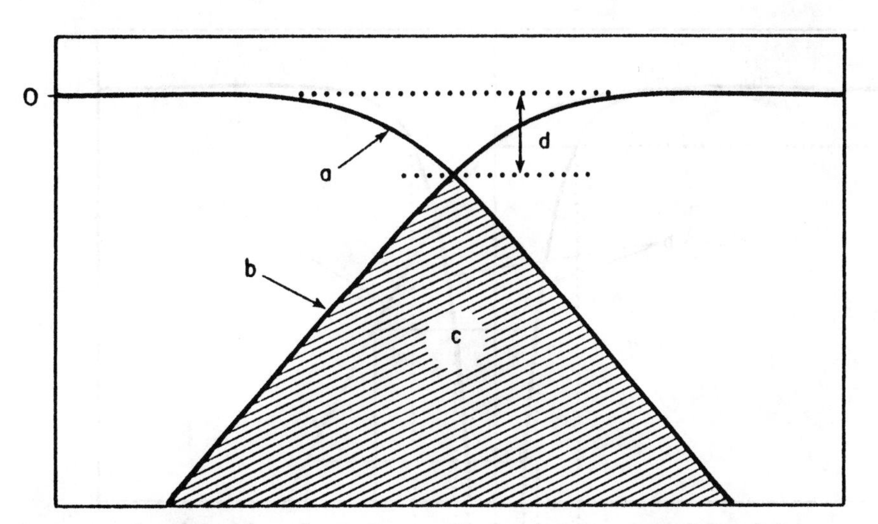

Figure 6.9 Centerband loss in a bandpass filter realized by cascading lowpass and highpass filters: (*a*) lowpass response, (*b*) highpass response, (*c*) passband of BPF, (*d*) centerband loss.

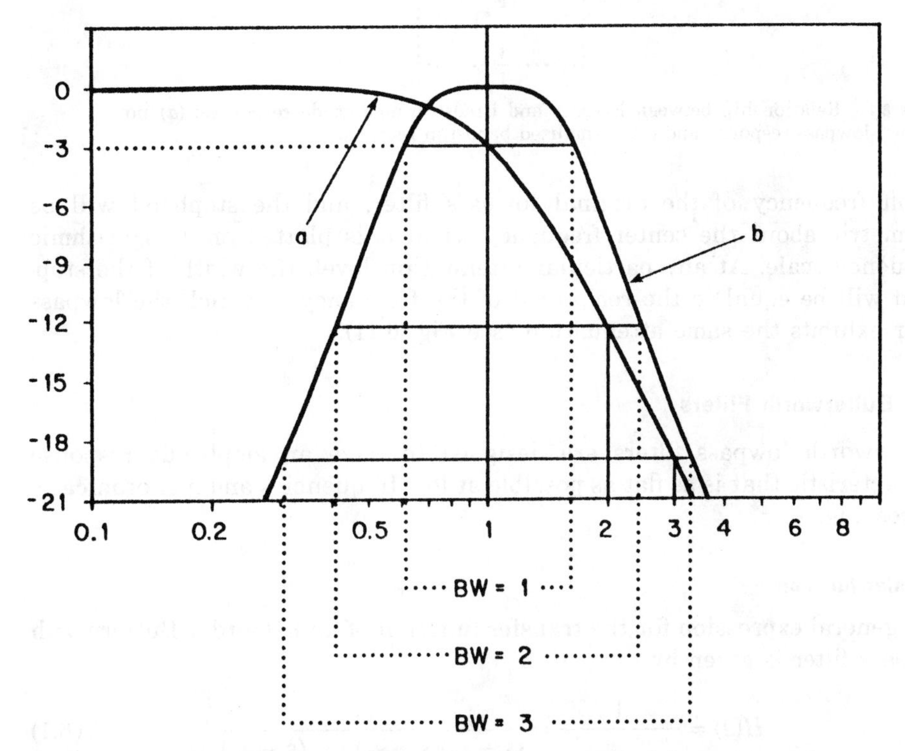

Figure 6.10 Relationship between lowpass and bandpass magnitude responses: (*a*) normalized lowpass response and (*b*) normalized bandpass response.

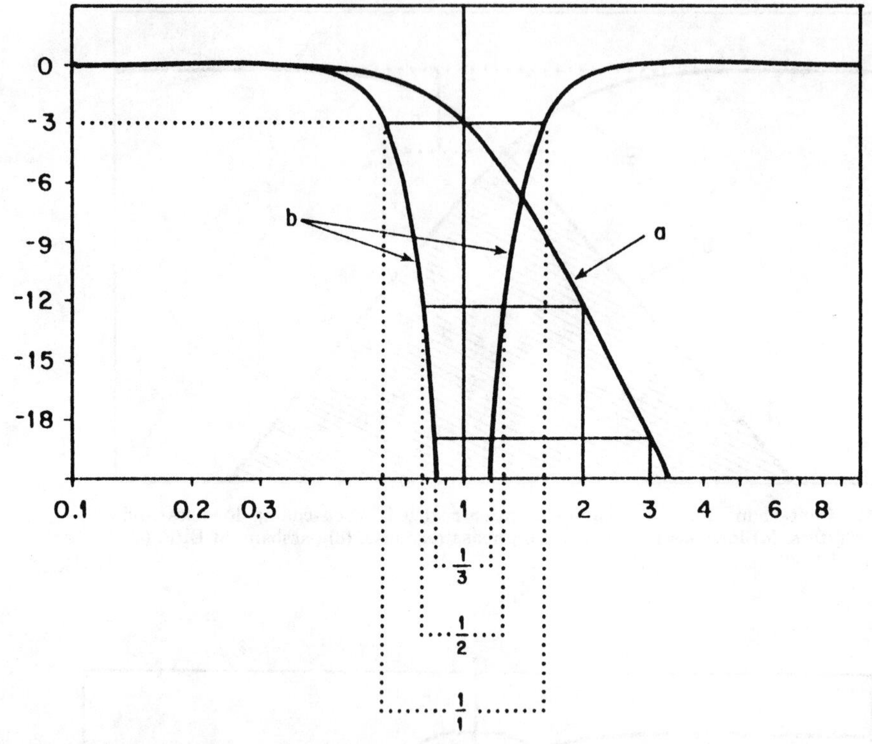

Figure 6.11 Relationship between lowpass and bandstop magnitude responses: (*a*) normalized lowpass response and (*b*) normalized bandstop response.

cutoff frequency of the original lowpass filter, and the stopband will be symmetric about the center frequency when it is plotted on a logarithmic frequency scale. At any particular attenuation level, the width of the stopband will be equal to the reciprocal of the frequency at which the lowpass filter exhibits the same attenuation (see Fig. 6.11).

6.2 Butterworth Filters

Butterworth lowpass filters are designed to have an amplitude response characteristic that is as flat as possible at low frequencies and monotonically decreasing.

Transfer function

The general expression for the transfer function of an nth-order Butterworth lowpass filter is given by

$$H(s) = \frac{1}{\prod_{i=1}^{n} (s - s_i)} = \frac{1}{(s - s_1)(s - s_2) \cdots (s - s_n)} \tag{6.1}$$

where

$$s_i = e^{j\pi[(2i+n-1)/(2n)]} = \cos \pi \frac{2i+n-1}{2n} + j \sin \pi \frac{2i+n-1}{2n} \tag{6.2}$$

Example 6.1 Determine the transfer function for a lowpass third-order Butterworth filter.

solution The third-order transfer function will have the form

$$H(s) = \frac{1}{(s-s_1)(s-s_2)(s-s_3)}$$

The values for s_1, s_2, and s_3 are obtained from Eq. (6.2):

$$s_1 = \cos \frac{2\pi}{3} + j \sin \frac{2\pi}{3} = -0.5 + 0.866j$$

$$s_2 = e^{j\pi} = \cos \pi + j \sin \pi = -1$$

$$s_3 = \cos \frac{4\pi}{3} + j \sin \frac{4\pi}{3} = -0.5 - 0.866j$$

Thus,

$$H(s) = \frac{1}{(s+0.5-0.866j)(s+1)(s+0.5-0.866j)}$$

$$= \frac{1}{s^3 + 2s^2 + 2s + 1}$$

The form of Eq. (6.1) indicates that an nth-order Butterworth filter will always have n poles and no finite zeros. Also true, but not quite so obvious, is the fact that these poles lie at equally spaced points on the left half of a circle in the s plane. As shown in Fig. 6.12 for the third-order case, any

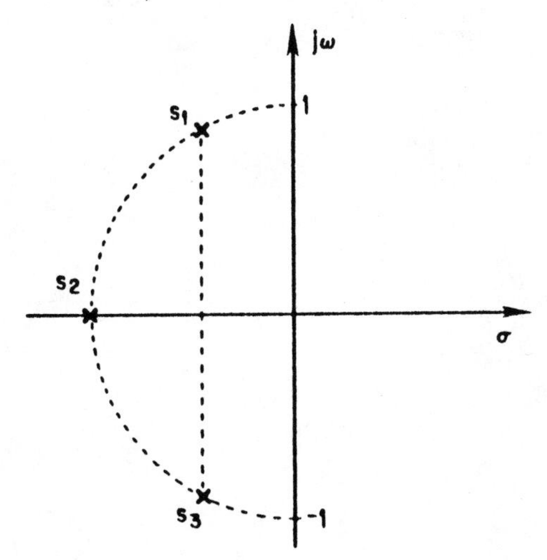

Figure 6.12 Pole locations for a third-order Butterworth LPF.

odd-order Butterworth lowpass filter (LPF) will have one real pole at $s = -1$, and all remaining poles will occur in complex conjugate pairs. As shown in Fig. 6.13 for the fourth-order case, the poles of any even-order Butterworth LPF will all occur in complex conjugate pairs. Pole values for orders 2 through 8 are listed in Table 6.1.

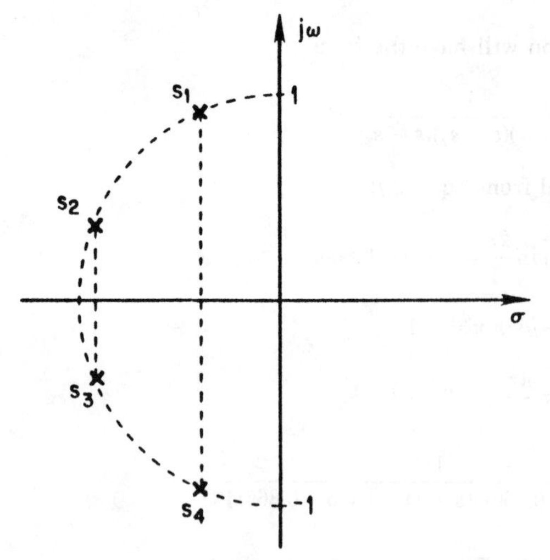

Figure 6.13 Pole locations for a fourth-order Butterworth LPF.

TABLE 6.1 Poles of Lowpass Butterworth Filters

n	Pole values
2	$-0.707107 \pm 0.707107j$
3	-1.0
	$-0.5 \pm 0.866025j$
4	$-0.382683 \pm 0.923880j$
	$-0.923880 \pm 0.382683j$
5	-1.0
	$-0.809017 \pm 0.587785j$
	$-0.309017 \pm 0.951057j$
6	$-0.258819 \pm 0.965926j$
	$-0.707107 \pm 0.707107j$
	$-0.965926 \pm 0.258819j$
7	-1.0
	$-0.900969 \pm 0.433884j$
	$-0.623490 \pm 0.781831j$
	$-0.222521 \pm 0.974928j$
8	$-0.195090 \pm 0.980785j$
	$-0.555570 \pm 0.831470j$
	$-0.831470 \pm 0.555570j$
	$-0.980785 \pm 0.195090j$

Frequency response

The C function **ButterworthFreqResponse( )** for generating Butterworth frequency response data is provided in Listing 6.2. Figures 6.14 through 6.16 show, respectively, the passband magnitude response, the stopband magnitude response, and the phase response for Butterworth filters of various orders. These plots are normalized for a cutoff frequency of 1 Hz. To denormalize them, simply multiply the frequency axis by the desired cutoff frequency f_c.

```
/************************************/
/*                                  */
/*   Listing 6.2                    */
/*                                  */
/*   ButterworthFreqResponse()      */
/*                                  */
/************************************/
#include <math.h>
#include <stdio.h>
#include "globDefs.h"
#include "protos.h"

void ButterworthFreqResponse( int order,
                              real frequency,
                              real *magnitude,
                              real *phase)
{
struct complex pole, s, numer, denom, transferFunction;
real x;
int k;
numer = cmplx(1.0,0.0);
denom = cmplx(1.0,0.0);

s = cmplx(0.0, frequency);
for( k=1; k<=order; k++)
    {
    x = PI * ((double)(order + (2*k)-1)) / (double)(2*order);
    pole = cmplx( cos(x), sin(x));
    denom = cMult(denom, cSub(s,pole));
    }
transferFunction = cDiv(numer, denom);
*magnitude = 20.0 * log10(cAbs(transferFunction));
*phase = 180.0 * arg(transferFunction) / PI;
return;
}
```

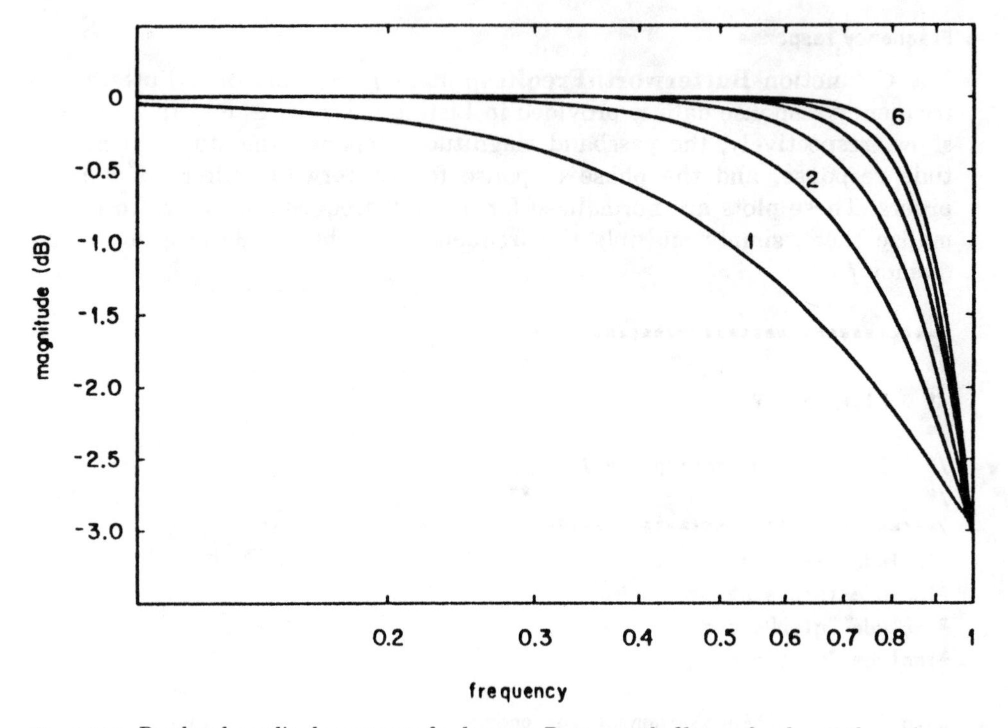

Figure 6.14 Passband amplitude response for lowpass Butterworth filters of orders 1 through 6.

Example 6.2 Use Figs. 6.15 and 6.16 to determine the magnitude and phase response at 800 Hz of a sixth-order Butterworth lowpass filter having a cutoff frequency of 400 Hz.

solution By setting $f_c = 400$, the $n = 6$ response of Fig. 6.15 is denormalized to obtain the response shown in Fig. 6.17. This plot shows that the magnitude at 800 Hz is approximately -36 dB. The corresponding response calculated by **ButterworthFreq-Response( )** is -36.12466 dB. Likewise, the $n = 6$ response of Fig. 6.16 is denormalized to obtain the response shown in Fig. 6.18. This plot shows that the phase response at 800 Hz is approximately $-425°$. The corresponding value calculated by **Butterworth-FreqResponse( )** is $-65.474°$ which "unwraps" to $-425.474°$.

Determination of minimum filter order

Usually, in the real world, the order of the desired filter is not given as in Example 6.2, but instead the order must be chosen based on the required performance of the filter. For lowpass Butterworth filters, the minimum order n that will ensure a magnitude of A_1 or lower at all frequencies ω_1 and above can be obtained by using

$$n = \frac{\log(10^{-A_1/10} - 1)}{2\log(\omega_1/\omega_c)} \tag{6.3}$$

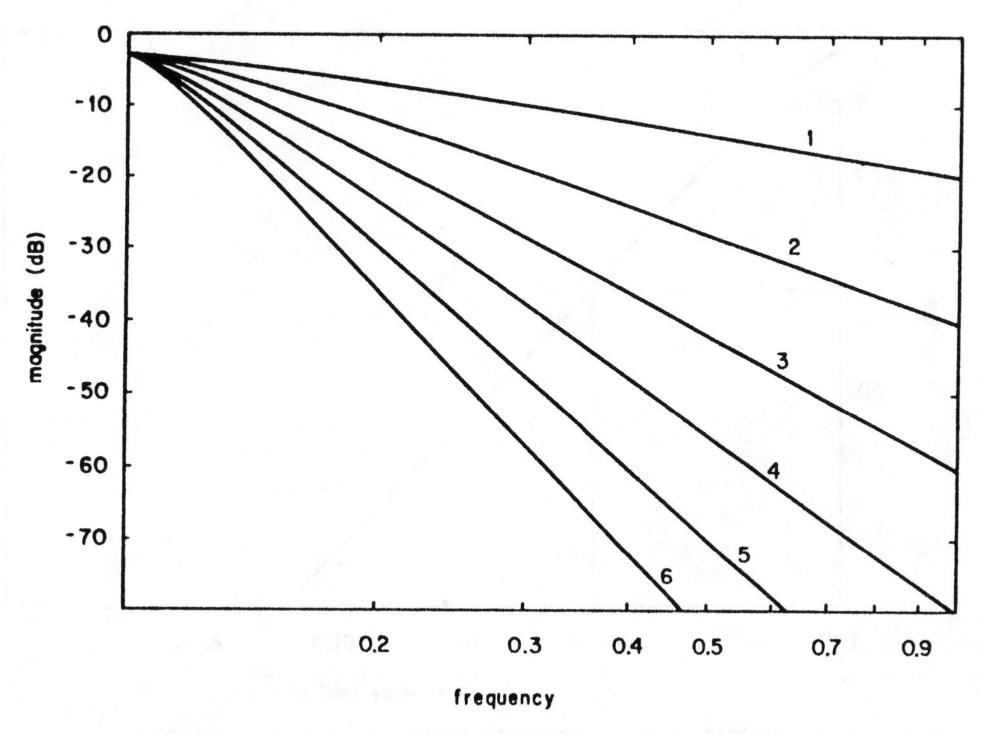

Figure 6.15 Stopband amplitude response for lowpass Butterworth filters of orders 1 through 6.

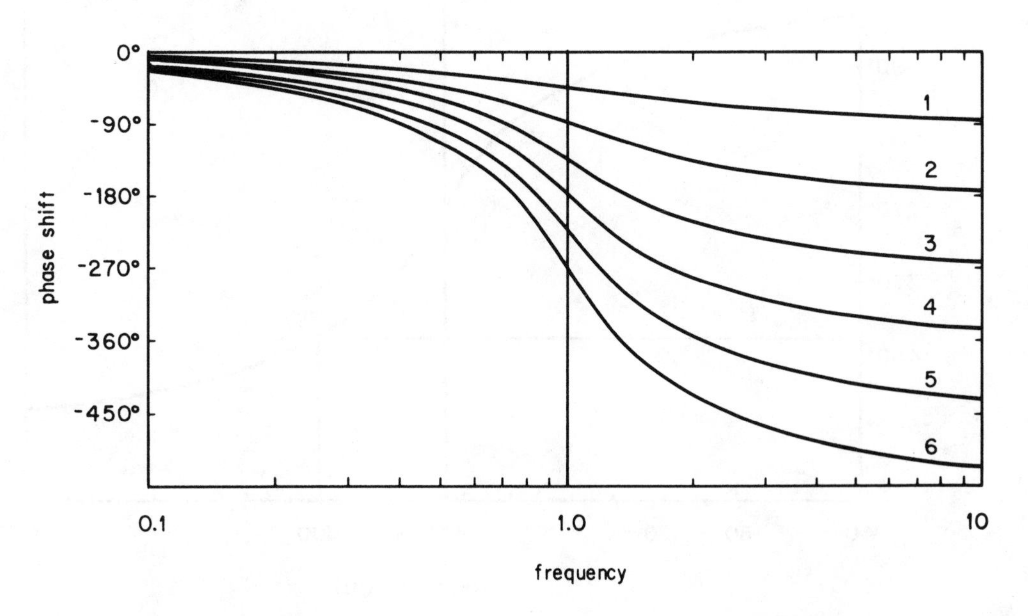

Figure 6.16 Phase response for lowpass Butterworth filters of orders 1 through 6.

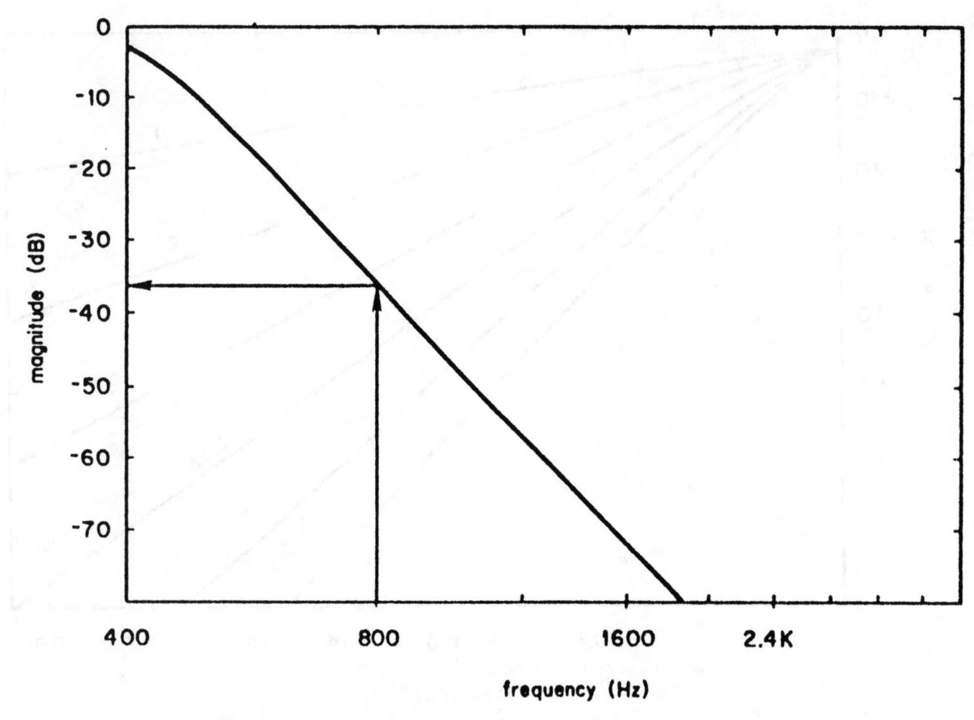

Figure 6.17 Denormalized amplitude response for Example 6.2.

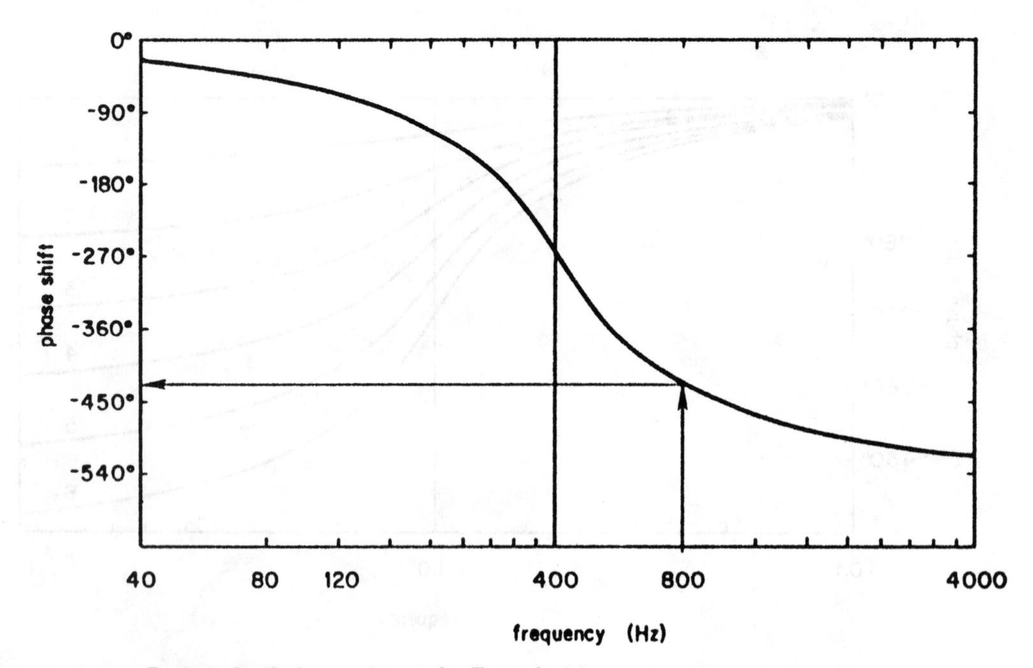

Figure 6.18 Denormalized phase response for Example 6.2.

where ω_c = 3-dB frequency and ω_1 = frequency at which the magnitude response first falls below A_1. (*Note*: The value of A_1 is assumed to be in decibels. The value will be negative, thus canceling the minus sign in the numerator exponent.)

Impulse response

To obtain the impulse response for an nth-order Butterworth filter, we need to take the inverse Laplace transform of the transfer function. Application of the Heaviside expansion to Eq. (6.1) produces

$$h(t) = \mathscr{L}^{-1}[h(t)] = \sum_{r=1}^{n} K_r e^{s_r t} \tag{6.4}$$

where
$$K_r = \left. \frac{s - s_r}{(s - s_1)(s - s_2) \cdots (s - s_n)} \right|_{s = s_r}$$

The values of both K_r and s_r are in general complex, but for the lowpass Butterworth case all the complex pole values occur in complex conjugate pairs. When the order n is even, this will allow Eq. (6.4) to be put in the form

$$h(t) = \sum_{r=1}^{n/2} (2\,\mathrm{Re}\{K_r\} e^{\sigma_r t} \cos \omega_r t - 2\,\mathrm{Im}\{K_r\} e^{\sigma_r t} \sin \omega_r t) \tag{6.5}$$

where $s_r = \sigma_r + j\omega_r$ and the roots s_r are numbered such that for $r = 1, 2, \ldots, n/2$ the s_r lie in the same quadrant of the s plane. [This last restriction prevents two members of the same complex conjugate pair from being used independently in evaluation of (6.5).] When the order n is odd, Eq. (6.4) can be put into the form

$$h(t) = Ke^{-t} + \sum_{r=1}^{(n-1)/2} (2\,\mathrm{Re}\{K_r\} e^{\sigma_r t} \cos \omega_r t - 2\,\mathrm{Im}\{K_r\} e^{\sigma_r t} \sin \omega_r t) \tag{6.6}$$

where no two of the roots s_r, $r = 1, 2, \ldots, (n-1)/2$, form a complex conjugate pair. Equations (6.5) and (6.6) form the basis for the C routine provided in Listing 6.3. This routine was used to generate the impulse responses for lowpass Butterworth filters shown in Figs. 6.19 and 6.20. These responses are normalized for lowpass filters having a cutoff frequency equal to 1 rad/s. To denormalize the response, divide the time axis by the desired cutoff frequency $\omega_c = 2\pi f_c$ and multiply the time axis by the same factor.

Example 6.3 Determine the instantaneous amplitude of the output 1.6 milliseconds (ms) after a unit impulse is applied to the input of a fifth-order Butterworth LPF having $f_c = 250$ Hz.

solution The $n = 5$ response of Fig. 6.20 is denormalized as shown in Fig. 6.21. This plot shows that the response amplitude at $t = 1.6$ ms is approximately 378.

```
/*************************************************/
/*                                               */
/*   Listing 6.3                                 */
/*                                               */
/*   ButterworthImpulseResponse()                */
/*                                               */
/*************************************************/
#include <math.h>
#include <stdio.h>
#include "globDefs.h"
#include "protos.h"

void ButterworthImpulseResponse(  int order,
                                  real delta_t,
                                  int npts,
                                  real yval[])
{
real L, M, x, R, I, LT, MT, cosPart, sinPart, h_of_t;
real K, sigma, omega, t;
int ix, r, ii, iii;
real ymax, ymin;

for( ix=0; ix <= npts; ix++)
    {
    printf("%d/n",ix);
    h_of_t = 0.0;
    t = delta_t * ix;
    for( r=1; r <= (order>>1); r++)
        {
        x = PI * (double)(order + (2*r)-1) / (double)(2*order);
        sigma = cos(x);
        omega = sin(x);
/*  Compute Lr and Mr    */

L = 1.0;
M = 0.0;
for( ii=1; ii<=order; ii++)
    {
    if( ii == r ) continue;
    x = PI * (double)(order + (2*ii)-1) / (double)(2*order);
    R = sigma - cos(x);
    I = omega - sin(x);
LT = L*R - M*I;
MT = L*I + R*M;
L = LT;
```

```
            M = MT;
            }
        L = LT / (LT*LT + MT*MT);
        M = -MT /(LT*LT + MT*MT);
        cosPart = 2.0 * L * exp(sigma*t) * cos(omega*t);
        sinPart = 2.0 * M * exp(sigma*t) * sin(omega*t);

        h_of_t = h_of_t + cosPart - sinPart;
        }
    if( (order%2) == 0)
        {
        yval[ix] = h_of_t;
        if( (real) h_of_t > ymax) ymax = h_of_t;
        if( (real) h_of_t < ymin) ymin = h_of_t;
        continue;
        }
    /* compute the real exponential component for odd-order responses */

        K = 1.0;
        L = 1.0;
        M = 0.0;
        r = (order+1)/2;
        x = PI * (double)(order + (2*r)-1) / (double)(2*order);
        sigma = cos(x);
        omega = sin(x);
        for( iii=1; iii<=order; iii++)
            {
            if( iii == r) continue;
            x = PI * (double)(order + (2*iii)-1) / (double)(2*order);
            R = sigma - cos(x);
            I = omega - sin(x);

            LT = L*R - M*I;
            MT = L*I + R*M;
            L = LT;
            M = MT;
            }
        K = LT / (LT*LT + MT*MT);
        h_of_t = h_of_t + K * exp(-t);
        yval[ix] = h_of_t;
        if( (real) h_of_t > ymax) ymax = h_of_t;
        if( (real) h_of_t < ymin) ymin = h_of_t;
        }
    return;
    }
```

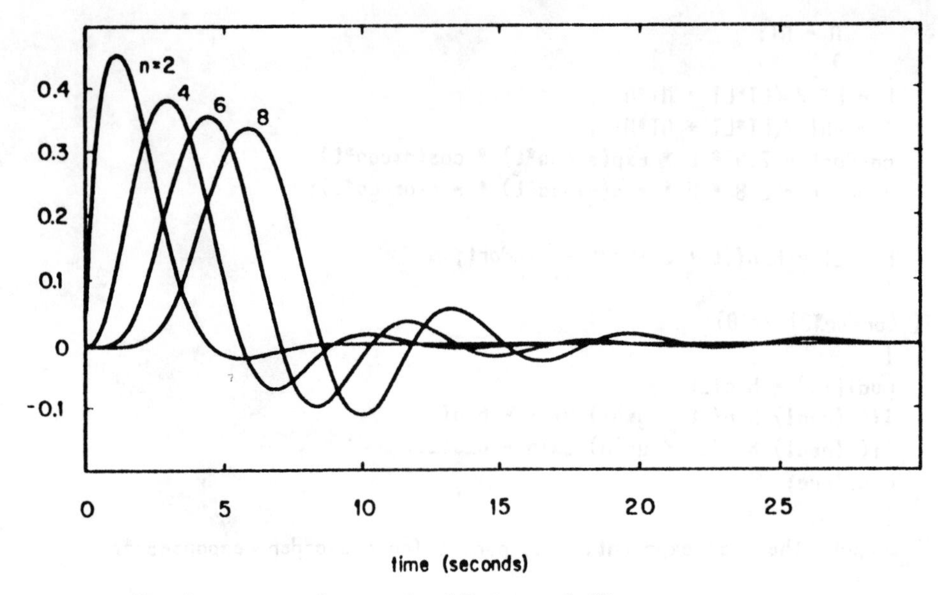

Figure 6.19 Impulse response of even-ordered Butterworth filters.

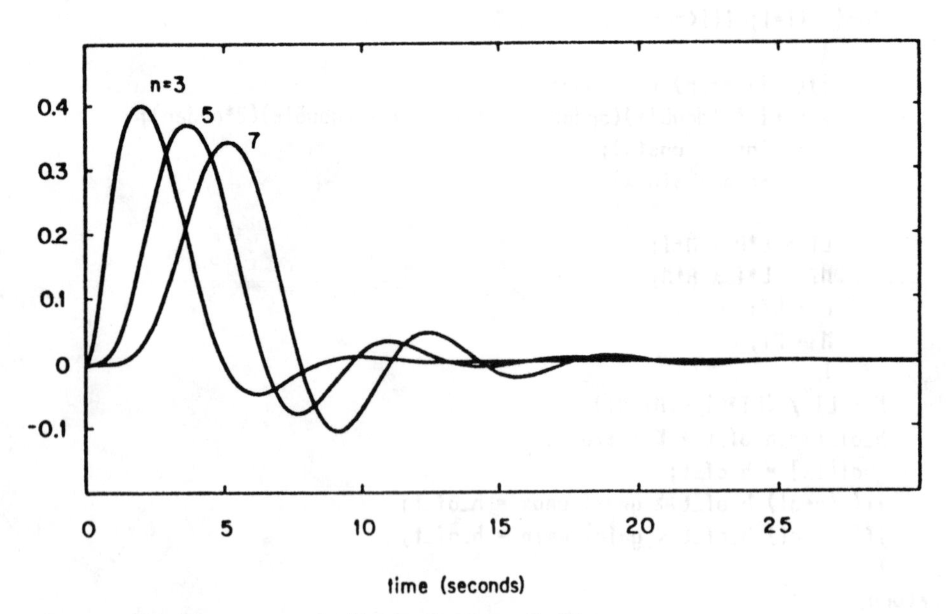

Figure 6.20 Impulse response of odd-order Butterworth filters.

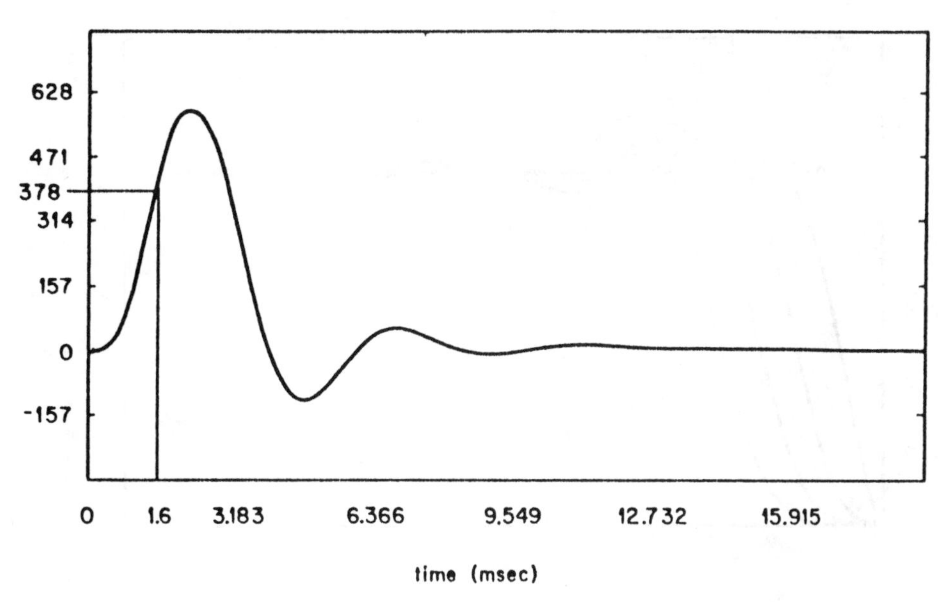

Figure 6.21 Denormalized impulse response for Example 6.3.

Step response

The step response can be obtained by integrating the impulse response. Step responses for lowpass Butterworth filters are shown in Figs. 6.22 and 6.23. These responses are normalized for lowpass filters having a cutoff frequency equal to 1 rad/s. To denormalize the response, divide the time axis by the desired cutoff frequency $\omega_c = 2\pi f_c$.

Example 6.4 Determine how long it will take for the step response of a third-order Butterworth LPF ($f_c = 4$ kHz) to first reach 100 percent of its final value.

solution By setting $\omega_c = 2\pi f_c = 8000\pi = 25{,}132.7$, the $n = 3$ response of Fig. 6.23 is denormalized to obtain the response shown in Fig. 6.24. This plot indicates that the step response first reaches a value of 1 in approximately 150 microseconds (μs).

6.3 Chebyshev Filters

Chebyshev filters are designed to have an amplitude response characteristic that has a relatively sharp transition from the passband to the stopband. This sharpness is accomplished at the expense of ripples that are introduced into the response. Specifically, Chebyshev filters are obtained as an equiripple approximation to the passband of an ideal lowpass filter. This results in a filter characteristic for which

$$|H(j\omega)|^2 = \frac{1}{1 + \epsilon^2 T_n^2(\omega)} \tag{6.7}$$

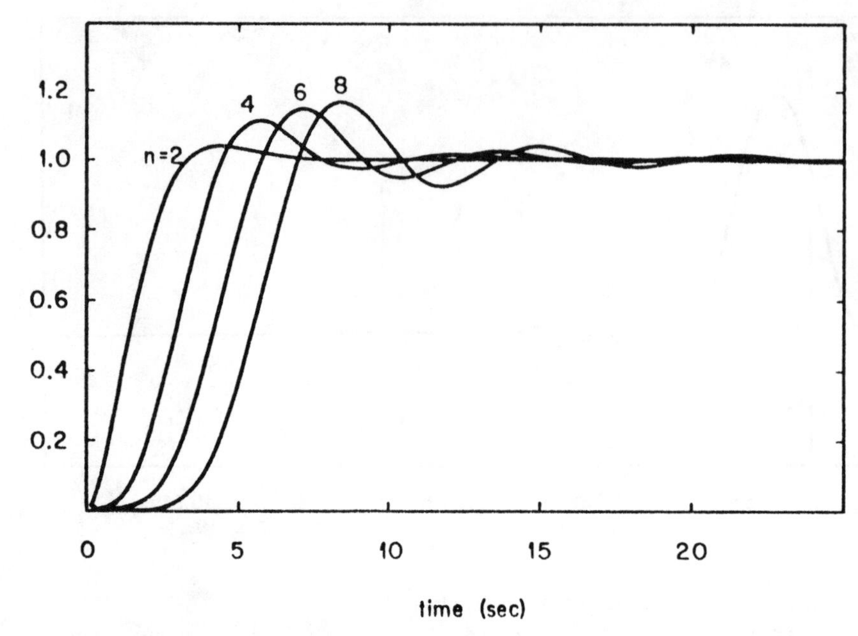

Figure 6.22 Step response of even-order lowpass Butterworth filters.

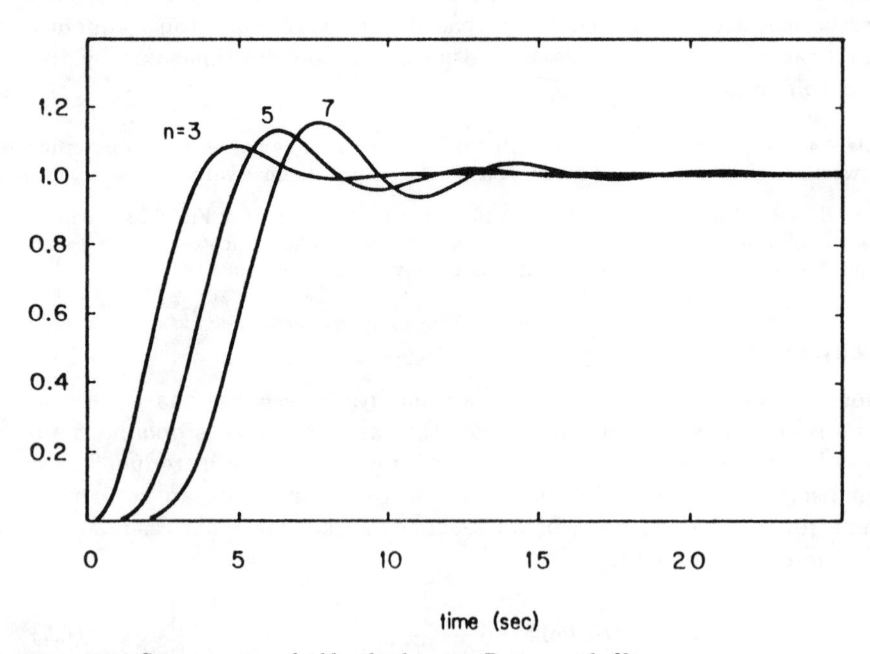

Figure 6.23 Step response of odd-order lowpass Butterworth filters.

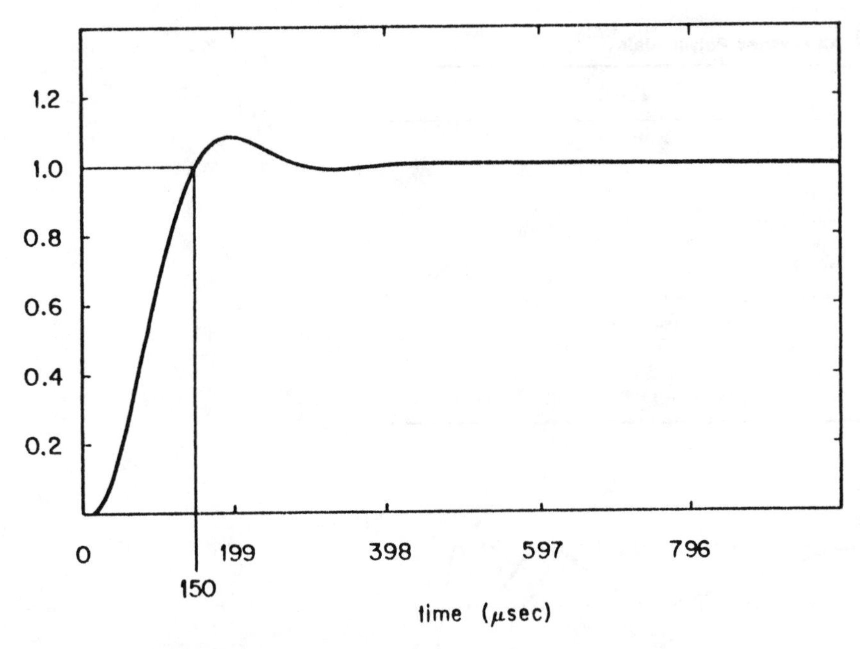

Figure 6.24 Denormalized step response for Example 6.4.

where $\epsilon^2 = 10^{r/10} - 1$
$T_n(\omega)$ = Chebyshev polynomial of order n
r = passband ripple, dB

Chebyshev polynomials are listed in Table 6.2.

Transfer function

The general shape of the Chebyshev magnitude response will be as shown in Fig. 6.25. This response can be normalized as in Fig. 6.26 so that the ripple bandwidth ω_r is equal to 1, or the response can be normalized as in Fig. 6.27 so that the 3-dB frequency ω_0 is equal to 1. Normalization based on the ripple bandwidth involves simpler calculations, but normalization based on the 3-dB point makes it easier to compare Chebyshev responses to those of other filter types.

The general expression for the transfer function of an nth-order Chebyshev lowpass filter is given by

$$H(s) = \frac{H_0}{\prod_{i=1}^{n} (s - s_i)} = \frac{H_0}{(s - s_1)(s - s_2) \cdots (s - s_n)} \tag{6.8}$$

TABLE 6.2 Chebyshev Polynomials

n	$T_n(\omega)$
0	1
1	ω
2	$2\omega^2 - 1$
3	$4\omega^3 - 3\omega$
4	$8\omega^4 - 8\omega^2 + 1$
5	$16\omega^5 - 20\omega^3 + 5\omega$
6	$32\omega^6 - 48\omega^4 + 18\omega^2 - 1$
7	$64\omega^7 - 112\omega^5 + 56\omega^3 - 7\omega$
8	$128\omega^8 - 256\omega^6 + 160\omega^4 - 32\omega^2 + 1$
9	$256\omega^9 - 576\omega^7 + 432\omega^5 - 120\omega^3 + 9\omega$
10	$512\omega^{10} - 1280\omega^8 + 1120\omega^6 - 400\omega^4 + 50\omega^2 + 1$

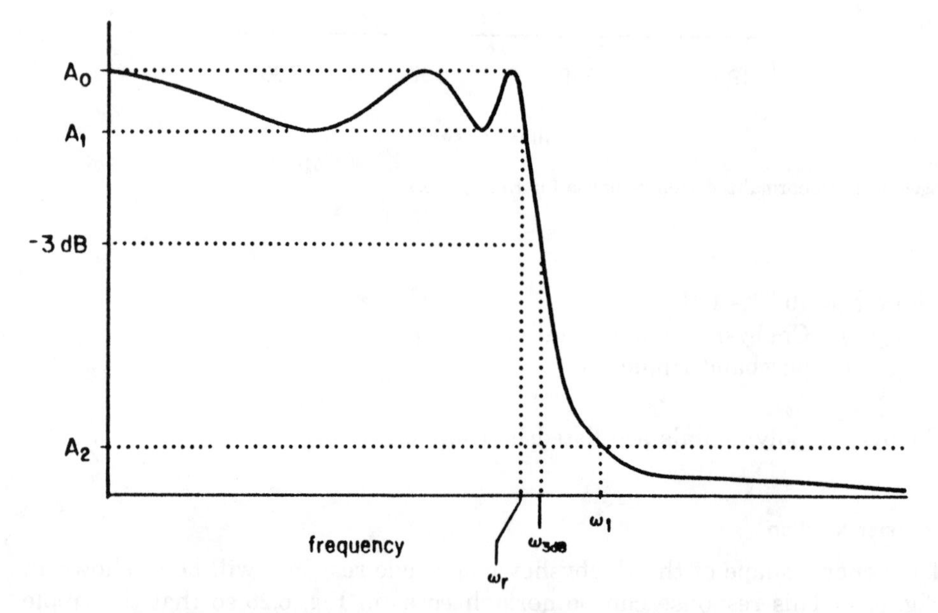

Figure 6.25 Magnitude response of a typical lowpass Chebyshev filter.

where

$$H_0 = \begin{cases} \prod_{i=1}^{n} (-s_i) & n \text{ odd} \\ 10^{r/20} \prod_{i=1}^{n} (-s_i) & n \text{ even} \end{cases} \tag{6.9}$$

$$s_i = \sigma_i + j\omega_i \tag{6.10}$$

$$\sigma_i = \frac{1/\gamma - \gamma}{2} \sin \frac{(2i-1)\pi}{2n} \tag{6.11}$$

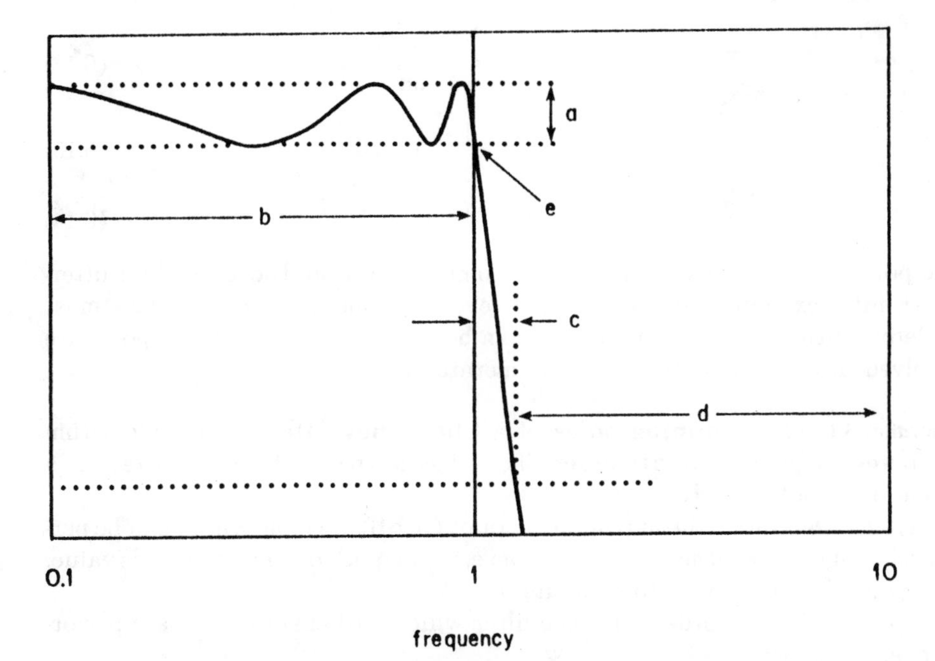

Figure 6.26 Chebyshev response normalized to have passband end at $\omega = 1$ rad/s. Features are (*a*) ripple limits, (*b*) passband, (*c*) transition band, (*d*) stopband, and (*e*) intersection of response and lower ripple limit at $\omega = 1$.

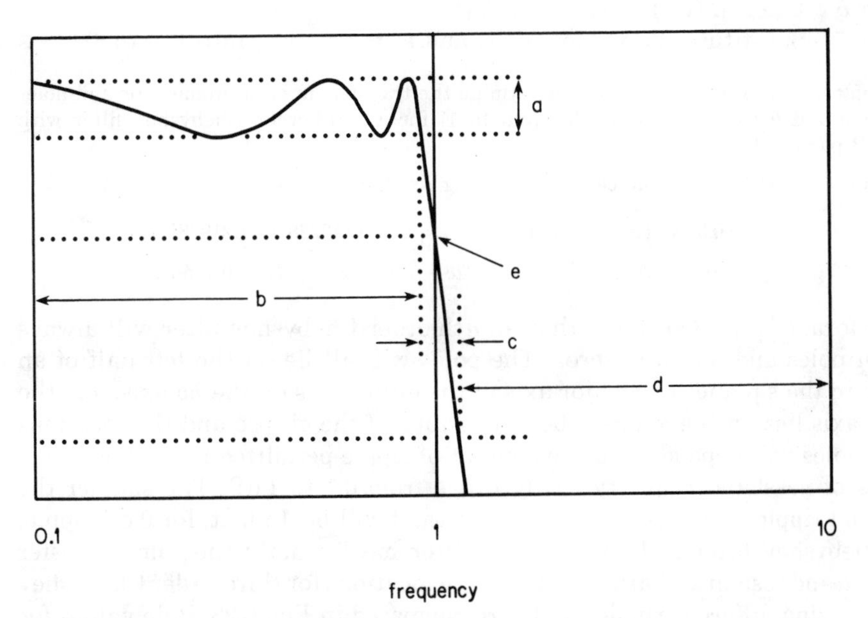

Figure 6.27 Chebyshev response normalized to have 3-dB point at $\omega = 1$ rad/s. Features are (*a*) ripple limits, (*b*) passband, (*c*) transition band, (*d*) stopband, and (*e*) response is 3 dB down at $\omega = 1$.

$$\omega_i = \frac{1/\gamma + \gamma}{2} \cos \frac{(2i-1)\pi}{2n} \tag{6.12}$$

$$\gamma = \left(\frac{1 + \sqrt{1 + \epsilon^2}}{\epsilon} \right)^{1/n} \tag{6.13}$$

$$\epsilon = \sqrt{10^{r/10} - 1} \tag{6.14}$$

The pole formulas are somewhat more complicated than those for the Butterworth filter examined in Sec. 6.2, and several parameters—ϵ, γ, and r—must be determined before the pole values can be calculated. Also all the poles are involved in the calculation of the numerator H_0.

Algorithm 6.1 **(Determining poles of a Chebyshev filter).** This algorithm computes the poles of an nth-order Chebyshev lowpass filter normalized for a ripple bandwidth of 1 Hz.

Step 1. Determine the maximum amount (in dB) of ripple which can be permitted in the passband magnitude response. Set r equal to or less than this value.

Step 2. Use Eq. (6.14) to compute ϵ.

Step 3. Select an order n for the filter which will ensure adequate performance.

Step 4. Use Eq. (6.13) to compute γ.

Step 5. For $i = 1, 2, \ldots, n$, use Eqs. (6.11) and (6.12) to compute the real part σ_i and imaginary part ω_i of each pole.

Step 6. Use Eq. (6.9) to compute H_0.

Step 7. Substitute the values of H_0 and s_1 through s_n into Eq. (6.8). ■

Example 6.5 Use Algorithm 6.1 to determine the transfer function numerator and poles (normalized for ripple bandwidth equal to 1) for a third-order Chebyshev filter with 0.5-dB passband ripple.

solution Algorithm 6.1 produces the following results:

$$\epsilon = 0.349311 \qquad \gamma = 1.806477 \qquad s_1 = -0.313228 + 1.021928j$$

$$s_2 = -0.626457 \qquad s_3 = -0.313228 - 1.021928j \qquad H_0 = 0.715695$$

The form of Eq. (6.8) shows that an nth-order Chebyshev filter will always have n poles and no finite zeros. The poles will all lie on the left half of an ellipse in the s plane. The major axis of the ellipse lies on the $j\omega$ axis, and the minor axis lies on the σ axis. The dimensions of the ellipse and the locations of the poles will depend upon the amount of ripple permitted in the passband. Values of passband ripple typically range from 0.1 to 1 dB. The smaller the passband ripple, the wider the transition band will be. In fact, for 0-dB ripple, the Chebyshev filter and Butterworth filter have exactly the same transfer function and response characteristics. Pole locations for third-order Chebyshev filters having different ripple limits are compared in Fig. 6.28. Pole values for ripple limits of 0.1, 0.5, and 1 dB are listed in Tables 6.3, 6.4, and 6.5 for orders 2 through 8.

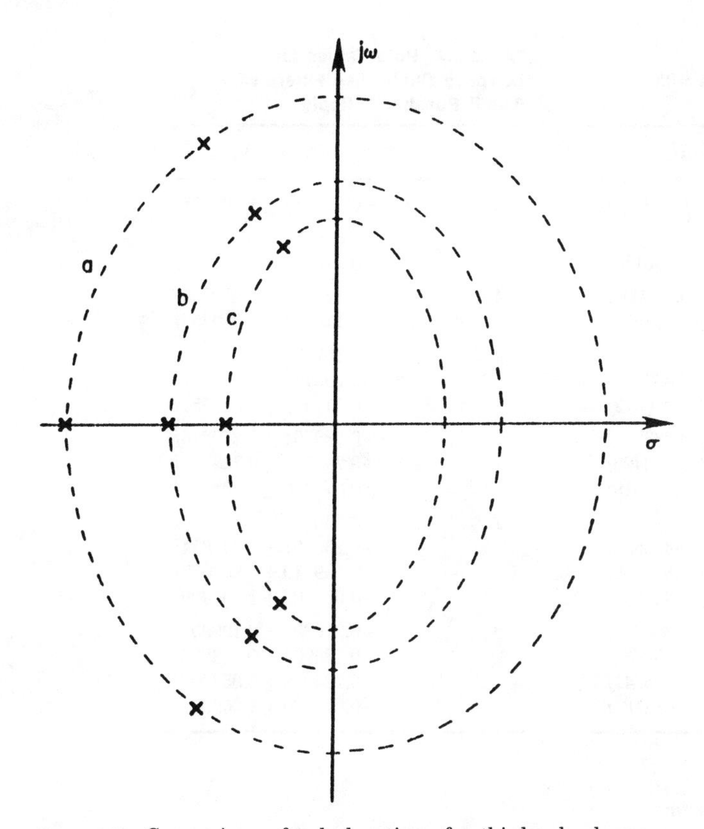

Figure 6.28 Comparison of pole locations for third-order lowpass Chebyshev filters with different amounts of passband ripple: (*a*) 0.01 dB, (*b*) 0.1 dB, and (*c*) 0.5 dB.

All the transfer functions and pole values presented so far are for filters normalized to have a ripple bandwidth of 1. Algorithm 6.2 can be used to renormalize the transfer function to have a 3-dB frequency of 1.

Algorithm 6.2 (Renormalizing Chebyshev LPF transfer functions). This algorithm assumes that ϵ, H_0, and the pole values s_i have been obtained for the transfer function having a ripple bandwidth of 1.

Step 1. Compute A, using

$$A = \frac{\cosh^{-1}(1/\epsilon)}{n} = \frac{1}{n} \log \frac{1 + \sqrt{1 - \epsilon^2}}{\epsilon}$$

Step 2. Using the value of A obtained in step 1, compute R as

$$R = \cosh A = \frac{e^A + e^{-A}}{2}$$

TABLE 6.3 Pole Values for Lowpass Chebyshev Filters with 0.1-dB Passband Ripple

n	Pole values
2	$-1.186178 \pm 1.380948j$
3	-0.969406
	$-0.484703 \pm 1.206155j$
4	$-0.637730 \pm 0.465000j$
	$-0.264156 \pm 1.122610j$
5	-0.538914
	$-0.435991 \pm 0.667707j$
	$-0.166534 \pm 1.080372j$
6	$-0.428041 \pm 0.283093j$
	$-0.313348 \pm 0.773426j$
	$-0.114693 \pm 1.056519j$
7	-0.376778
	$-0.339465 \pm 0.463659j$
	$-0.234917 \pm 0.835485j$
	$-0.083841 \pm 1.041833j$
8	$-0.321650 \pm 0.205314j$
	$-0.272682 \pm 0.584684j$
	$-0.182200 \pm 0.875041j$
	$-0.063980 \pm 1.032181j$

TABLE 6.4 Pole Values for Lowpass Chebyshev Filters with 0.5-dB Passband Ripple

n	Pole values
2	$-0.712812 \pm 1.00402j$
3	-0.626457
	$-0.313228 \pm 1.021928j$
4	$-0.423340 \pm 0.420946j$
	$-0.175353 \pm 1.016253j$
5	-0.362320
	$-0.293123 \pm 0.625177j$
	$-0.111963 \pm 1.011557j$
6	$-0.289794 \pm 0.270216j$
	$-0.212144 \pm 0.738245j$
	$-0.077650 \pm 1.008461j$
7	-0.256170
	$-0.230801 \pm 0.447894j$
	$-0.159719 \pm 0.807077j$
	$-0.057003 \pm 1.006409j$
8	$-0.219293 \pm 0.199907j$
	$-0.185908 \pm 0.569288j$
	$-0.124219 \pm 0.852000j$
	$-0.043620 \pm 1.005002j$

TABLE 6.5 Pole Values for Lowpass Chebyshev Filters with 1.0-dB Passband Ripple

n	Pole values
2	$-0.548867 \pm 0.895129j$
3	-0.494171
	$-0.247085 \pm 0.965999j$
4	$-0.336870 \pm 0.407329j$
	$-0.139536 \pm 0.983379j$
5	-0.289493
	$-0.234205 \pm 0.611920j$
	$-0.089458 \pm 0.990107j$
6	$-0.232063 \pm 0.266184j$
	$-0.169882 \pm 0.727227j$
	$-0.062181 \pm 0.993411j$
7	-0.205414
	$-0.185072 \pm 0.442943j$
	$-0.128074 \pm 0.798156j$
	$-0.045709 \pm 0.995284j$
8	$-0.175998 \pm 0.198206j$
	$-0.149204 \pm 0.564444j$
	$-0.099695 \pm 0.844751j$
	$-0.035008 \pm 0.996451j$

(Table 6.6 lists R factors for various orders and ripple limits. If the required combination can be found in this table, steps 1 and 2 can be skipped.)

Step 3. Use R to compute $H_{3\,\mathrm{dB}}(s)$ as

$$H_{3\,\mathrm{dB}}(s) = \frac{H_0 / R^n}{\prod_{i=1}^{n} \left(s - \frac{s_i}{R} \right)}$$ ■

Frequency response

Figures 6.29 through 6.32 show the magnitude and phase responses for Chebyshev filters with passband ripple limits of 0.5 dB. For comparison purposes, Figs. 6.33 and 6.34 show Chebyshev passband responses for ripple limits of 0.1 and 1.0 dB. These plots are normalized for a cutoff frequency of 1 Hz. To denormalize them, simply multiply the frequency axis by the desired cutoff frequency f_c. The C function **ChebyshevFreqResponse( )**, used to generate the Chebyshev frequency response data, is provided in Listing 6.4. Note that this function incorporates Algorithms 6.1 and 6.2.

```
/************************************/
/*                                  */
/*   Listing 6.4                    */
/*                                  */
/*   ChebyshevFreqResponse()        */
/*                                  */
/************************************/
#include <math.h>
#define PI (double) 3.141592653589
void ChebyshevFreqResponse(  int order,
                             float ripple,
                             char normalizationType,
                             float frequency,
                             float *magnitude,
                             float *phase)

{
double A, gamma, epsilon, work;
double rp, ip, x, i, r, rpt, ipt;
double normalizedFrequency, hSubZero;
int k, ix;

epsilon = sqrt( -1.0 + pow( (double)10.0, (double)(ripple/10.0) ));
gamma = pow( (( 1.0 + sqrt( 1.0 + epsilon*epsilon))/epsilon),
               (double)(1.0/(float) order) );
```

```
if( normalizationType == '3' )
    {
    work = 1.0/epsilon;
    A = ( log( work + sqrt( work*work - 1.0) ) ) / order;
    normalizedFrequency = frequency * ( exp(A) + exp(-A))/2.0;
    }
else
    {
    normalizedFrequency = frequency;
    }

rp = 1.0;
ip = 0.0;
for( k=1; k<=order; k++)
    {
    x = (2*k-1) * PI / (2.0*order);
    i = 0.5 * (gamma + 1.0/gamma) * cos(x);
    r = -0.5 * (gamma - 1.0/gamma) * sin(x);
    rpt = ip * i - rp * r;
    ipt = -rp * i - r * ip;
    ip = ipt;
    rp = rpt;
    }
hSubZero = sqrt( ip*ip + rp*rp);
if( order%2 == 0 )
    {
    hSubZero = hSubZero / sqrt(1.0 + epsilon*epsilon);
    }
rp = 1.0;
ip = 0.0;
for( k=1; k<=order; k++)
    {
    x = (2*k-1)*PI/(2.0*order);
    i = 0.5 * (gamma + 1.0/gamma) * cos(x);
    r = -0.5 * (gamma - 1.0/gamma) * sin(x);
    rpt = ip*(i-normalizedFrequency) - rp*r;
    ipt = rp*(normalizedFrequency-i) - r*ip;
    ip=ipt;
    rp=rpt;
    }
*magnitude = 20.0 * log10(hSubZero/sqrt(ip*ip+rp*rp));
*phase = 180.0 * atan2( ip, rp) /PI;
return;
}
```

Impulse response

Impulse responses for lowpass Chebyshev filters with 0.5-dB ripple are shown in Fig. 6.35. The C routine **ChebyshevImpulseResponse( )**, used to generate the data for these plots, is provided in Listing 6.5. These responses are normalized for lowpass filters having a 3-dB frequency of 1 Hz. To denormalize the response, divide the time axis by the desired cutoff frequency f_c and multiply the amplitude axis by the same factor.

```
/***************************************************/
/*                                                 */
/*   Listing 6.5                                   */
/*                                                 */
/*   ChebyshevImpulseResponse()                    */
/*                                                 */
/***************************************************/
#include <math.h>
#define PI (double) 3.141592653589

void ChebyshevImpulseResponse(    int order,
                                  float ripple,
                                  char normalizationType,
                                  float delta_t,
                                  int npts,
                                  float yval[])

{
double m, p;
double A, gamma, epsilon, work, normFactor;
double rp, ip, x, i, r, rpt, ipt, ss;
double hSubZero, h_of_t, t, sigma, omega;
double K, L, M, LT, MT, I, R, cosPart, sinPart;
int k, ix, ii, iii, rrr;


epsilon = sqrt( -1.0 + pow( (double)10.0, (double)(ripple/10.0) ));

if( normalizationType == '3')
    {
    work = 1.0/epsilon;
    A = ( log( work + sqrt( work*work - 1.0) ) ) / order;
    normFactor = ( exp(A) + exp(-A))/2.0;
    }
```

```
else
    {
    normFactor = 1.0;
    }
gamma = pow( (( 1.0 + sqrt( 1.0 + epsilon*epsilon))/epsilon),
                (double)(1.0/(float) order) );
/*------------------------------*/
/*  compute H_zero               */
rp = 1.0;
ip = 0.0;

for( k=1; k<=order; k++)
    {
    x = (2*k-1) * PI / (float)(2*order);
    i = 0.5 * (gamma + 1.0/gamma) * cos(x)/normFactor;
    r = -0.5 * (gamma - 1.0/gamma) * sin(x)/normFactor;
    rpt = ip * i - rp * r;
    ipt = -rp * i - r * ip;
    ip = ipt;
    rp = rpt;
    }
hSubZero = sqrt( ip*ip + rp*rp);
if( order%2 == 0 )
    {
    hSubZero = hSubZero / sqrt(1.0 + epsilon*epsilon);
    }
printf("hSubZero = %f\n",hSubZero);
/*----------------------------------------*/
for( ix=0; ix<npts; ix++)
    {
    printf("%d\n",ix);
    h_of_t = 0.0;
    t = delta_t * ix;
    for( rrr=1; rrr <= (order >> 1); rrr++)
        {
        x = (2*rrr-1)*PI/(2.0*order);
        sigma = -0.5 * (gamma - 1.0/gamma) * sin(x)/normFactor;
        omega = 0.5 * (gamma + 1.0/gamma) * cos(x)/normFactor;
/*  compute Lr and Mr    */
L = 1;
M = 0;

for(ii=1; ii<=order; ii++)
    {
    if( ii == rrr) continue;
```

```
x = (2*ii-1) * PI /(float)(2*order);
R = sigma -(-0.5*(gamma -1.0/gamma))*sin(x) / normFactor;

I = omega -(0.5*(gamma +1.0/gamma))*cos(x) / normFactor;


    LT = L * R - M * I;
    MT = L * I + R * M;
    L = LT;
    M = MT;
    }
L = LT / (LT * LT + MT * MT);
M = -MT / (LT * LT + MT * MT);


cosPart = 2.0 * L * exp(sigma*t) * cos(omega*t);
sinPart = 2.0 * M * exp(sigma*t) * sin(omega*t);


    h_of_t = h_of_t + cosPart - sinPart;
    }
if( (order%2) == 0 )
    {
    yval[ix] = h_of_t * hSubZero;
    }

else
    {
    /*  compute the real exponential component    */
    /*  present in odd-order responses            */


K = 1;
L = 1;
M = 0;
rrr = (order+1) >> 1;

x = (2*rrr-1) * PI / (float)(2*order);


sigma = -0.5 * (gamma - 1.0/gamma) * sin(x) / normFactor;
omega = 0.5 * (gamma + 1.0/gamma) * cos(x) / normFactor;

for( iii=1; iii<= order; iii++)
    {
```

```
if(iii == rrr) continue;
x = (2*iii-1) * PI / (float)(2*order);
R = sigma -(-0.5*(gamma -1.0/gamma))*sin(x) / normFactor;
I = omega -(0.5*(gamma +1.0/gamma))*cos(x) / normFactor;

LT = L * R - M * I;
MT = L * I + R * M;
L = LT;
M = MT;
            }
        K = LT / (LT*LT + MT*MT);
        h_of_t = h_of_t + K * exp(sigma*t);
        yval[ix] = h_of_t * hSubZero;
        }
    }
return;
}
```

Step response

The step response can be obtained by integrating the impulse response. Step responses for lowpass Chebyshev filters with 0.5-dB ripple are shown in Fig. 6.36. These responses are normalized for lowpass filters having a cutoff frequency equal to 1 Hz. To denormalize the response, divide the time axis by the desired cutoff frequency f_c.

TABLE 6.6 Factors for Renormalizing Chebyshev Transfer Functions

	Order						
Ripple	2	3	4	5	6	7	8
0.1	1.94322	1.38899	1.21310	1.13472	1.09293	1.06800	1.05193
0.2	1.67427	1.28346	1.15635	1.09915	1.06852	1.05019	1.03835
0.3	1.53936	1.22906	1.12680	1.08055	1.05571	1.04083	1.03121
0.4	1.45249	1.19348	1.10736	1.06828	1.04725	1.03464	1.02649
0.5	1.38974	1.16749	1.09310	1.05926	1.04103	1.03009	1.02301
0.6	1.34127	1.14724	1.08196	1.05220	1.03616	1.02652	1.02028
0.7	1.30214	1.13078	1.07288	1.04644	1.03218	1.02361	1.01806
0.8	1.26955	1.11699	1.06526	1.04160	1.02883	1.02116	1.01618
0.9	1.24176	1.10517	1.05872	1.03745	1.02596	1.01905	1.01457
1.0	1.21763	1.09487	1.05300	1.03381	1.02344	1.01721	1.01316
1.1	1.19637	1.08576	1.04794	1.03060	1.02121	1.01557	1.01191
1.2	1.17741	1.07761	1.04341	1.02771	1.01922	1.01411	1.01079
1.3	1.16035	1.07025	1.03931	1.02510	1.01741	1.01278	1.00978
1.4	1.14486	1.06355	1.03558	1.02272	1.01576	1.01157	1.00886
1.5	1.13069	1.05740	1.03216	1.02054	1.01425	1.01046	1.00801

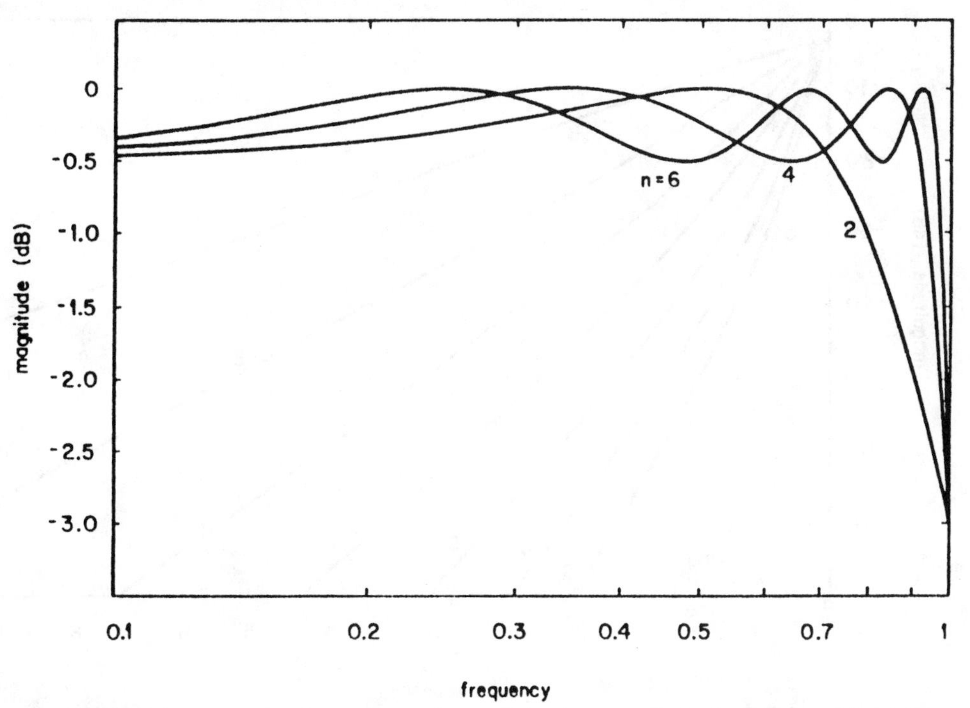

Figure 6.29 Passband magnitude response of even-order lowpass Chebyshev filters with 0.5-dB ripple.

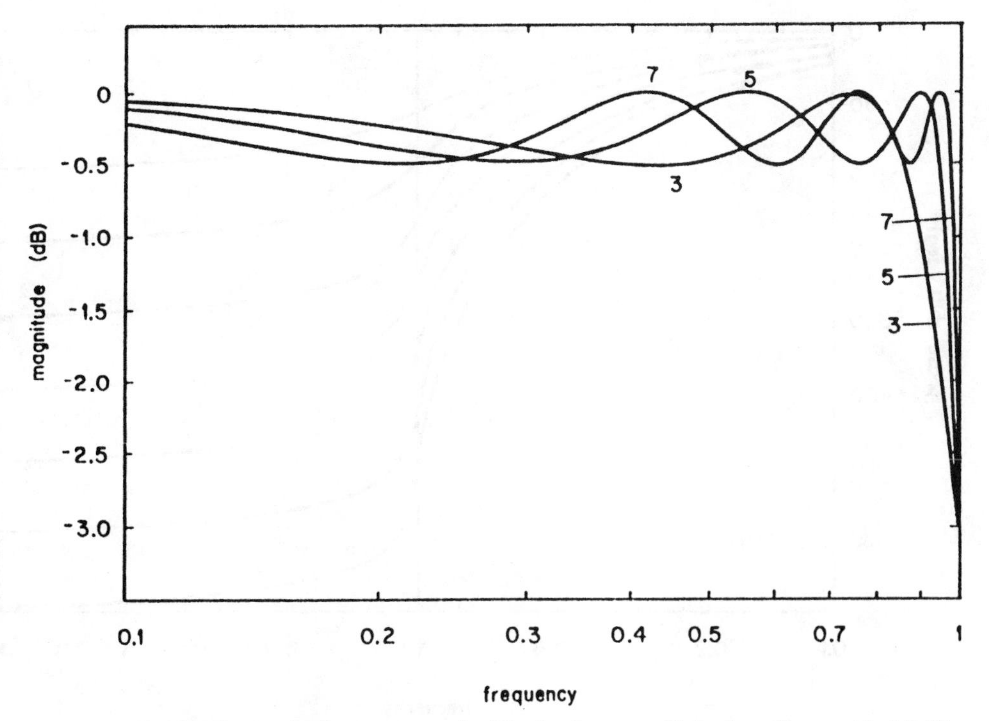

Figure 6.30 Passband magnitude response of odd-order lowpass Chebyshev filters with 0.5-dB ripple.

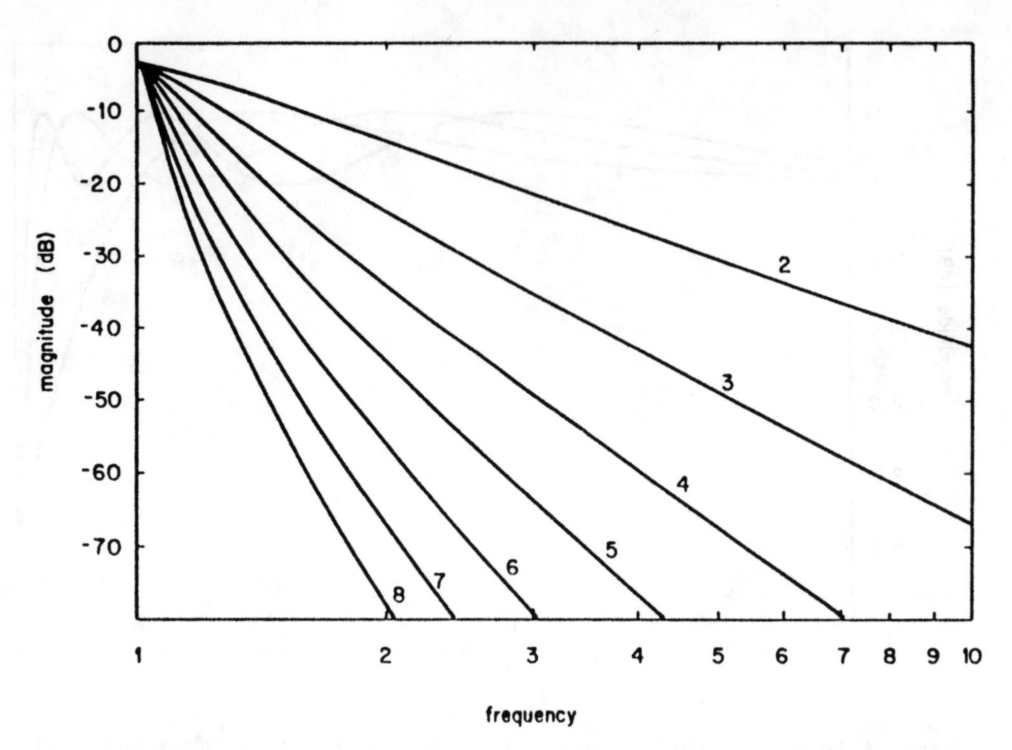

Figure 6.31 Stopband magnitude response of lowpass Chebyshev filters with 0.5-dB ripple.

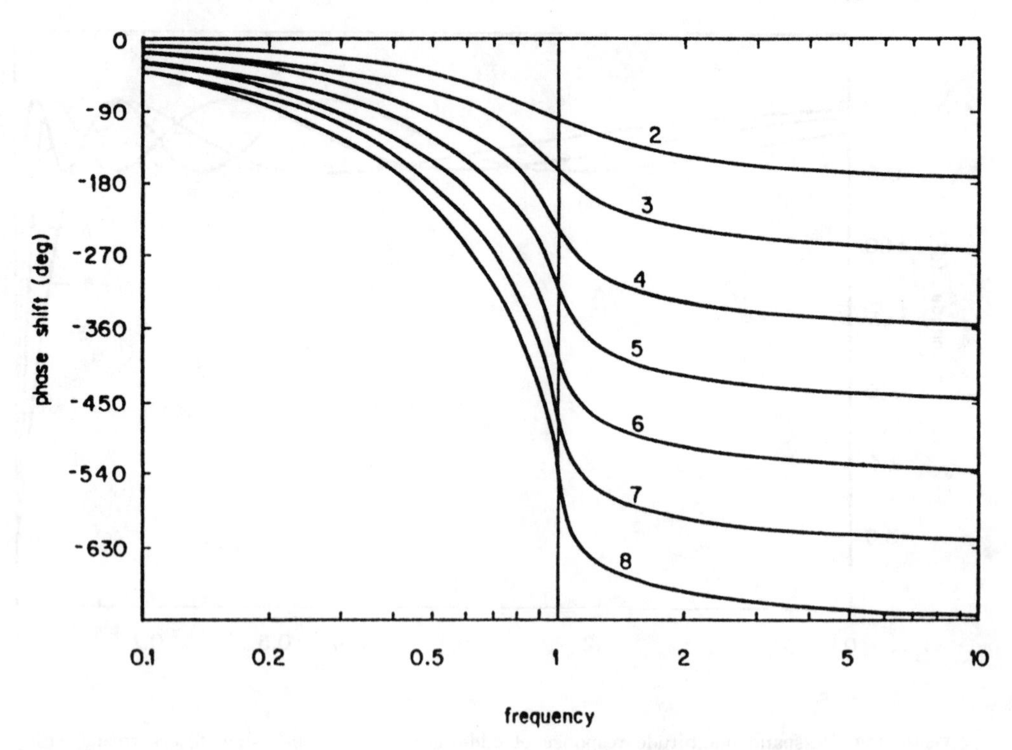

Figure 6.32 Phase response of lowpass Chebyshev filters with 0.5-dB passband ripple.

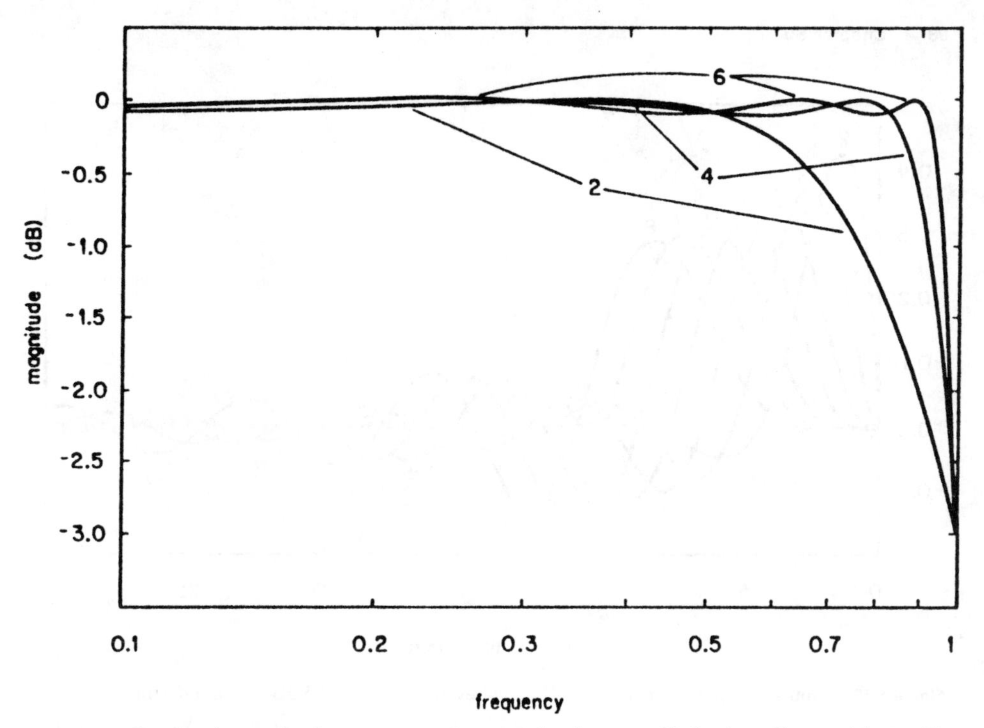

Figure 6.33 Passband magnitude response of even-order lowpass Chebyshev filters with 0.1-dB ripple.

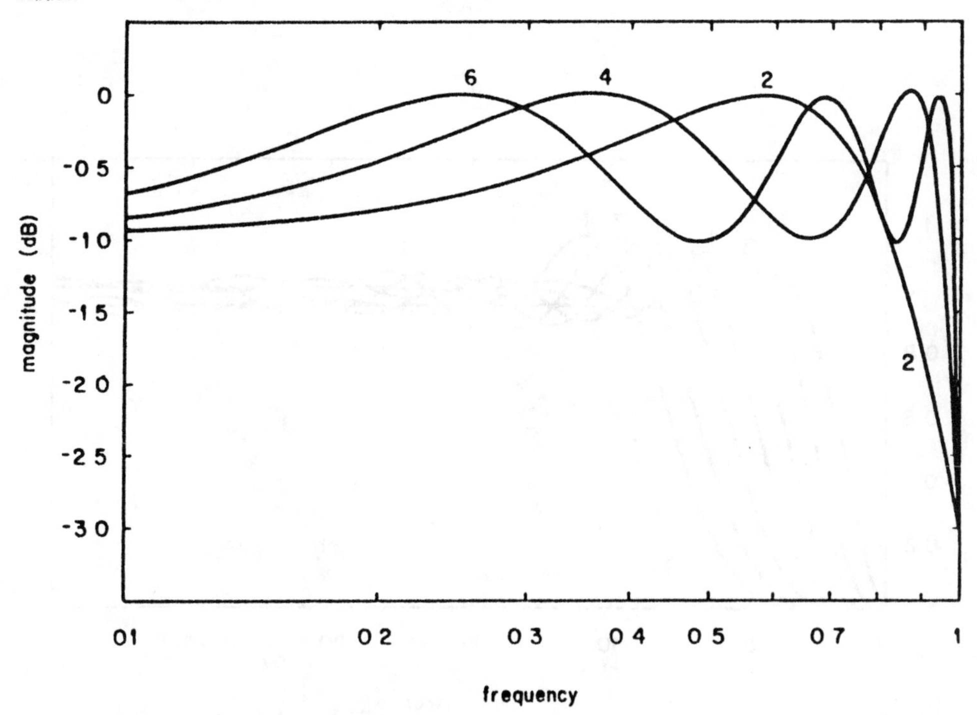

Figure 6.34 Passband magnitude response of even-order lowpass Chebyshev filters with 1.0-dB ripple.

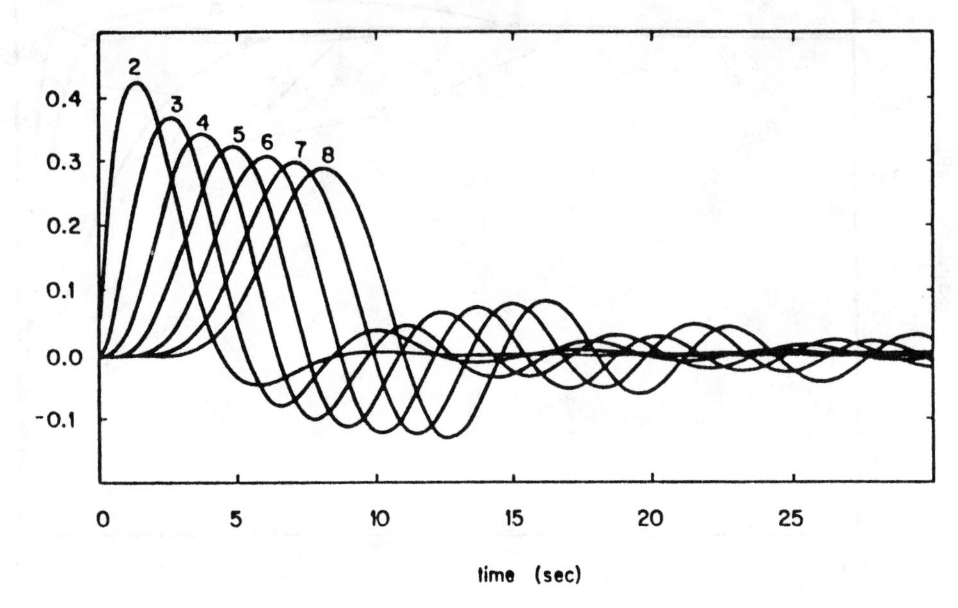

Figure 6.35 Impulse response of lowpass Chebyshev filters with 0.5-dB passband ripple.

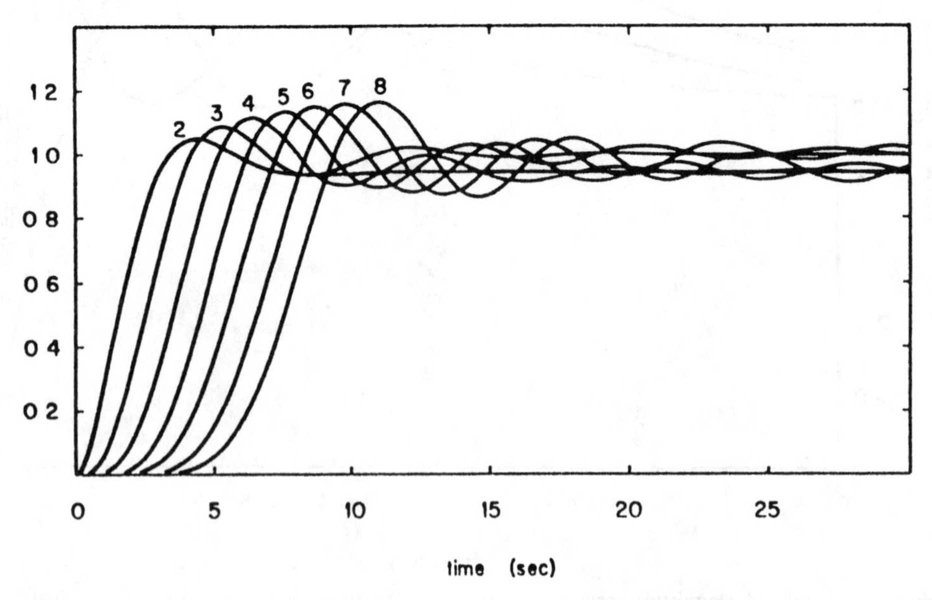

Figure 6.36 Step response of lowpass Chebyshev filters with 0.5-dB passband ripple.

6.4 Bessel Filters

Bessel filters are designed to have maximally flat group delay characteristics. As a consequence, there is no ringing in the impulse and step responses.

Transfer function

The general expression for the transfer function of an nth-order Bessel lowpass filter is given by

$$H(s) = \frac{b_0}{q_n(s)} \tag{6.15}$$

where

$$q_n(s) = \sum_{k=1}^{n} b_k s^k$$

$$b_k = \frac{(2n-k)!}{2^{n-k}k!(n-k)!}$$

The following recursion can be used to determine $q_n(s)$ from $q_{n-1}(s)$ and $q_{n-2}(s)$:

$$q_n = (2n-1)q_{n-1} + s^2 q_{n-2}$$

Table 6.7 lists $q_n(s)$ for $n = 2$ through $n = 8$. These values were generated by the C function **BesselCoefficients( )** provided in Listing 6.6. This function is used by other Bessel filter routines presented later in this section.

```
/************************************/
/*                                  */
/*   Listing 6.6                    */
/*                                  */
/*   BesselCoefficients()           */
/*                                  */
/************************************/

#include <math.h>
#include "globDefs.h"

void BesselCoefficients( int order,
                         char typeOfNormalization,
                         real coef[])

{
int i, N, index, indexM1, indexM2;
real B[3][MAXORDER];
real A, renorm[MAXORDER];
```

```
renorm[2] = 0.72675;
renorm[3] = 0.57145;
renorm[4] = 0.46946;
renorm[5] = 0.41322;
renorm[6] = 0.37038;
renorm[7] = 0.33898;
renorm[8] = 0.31546;
A = renorm[order];

index = 1;
indexM1 = 0;
indexM2 = 2;

for( i=0; i<(3*MAXORDER); i++) B[0][i] = 0;
B[0][0] = 1.0;
B[1][0] = 1.0;
B[1][1] = 1.0;
for( N=2; N<=order; N++)
    {
    index = (index+1)%3;
    indexM1 = (indexM1 + 1)%3;
    indexM2 = (indexM2 + 1)%3;

    for( i=0; i<N; i++)
        {
        B[index][i] = (2*N-1) * B[indexM1][i];
        }
    for( i=2; i<=N; i++)
        {
        B[index][i] = B[index][i] + B[indexM2][i-2];
        }
    }
if(typeOfNormalization == 'D')
    {
    for( i=0; i<=order; i++) coef[i] = B[index][i];
    }
else
    {
    for( i=0; i<=order; i++)
        {
        coef[i] = B[index][i] * pow(A, (order - i) );
        }
    }
return;
}
```

TABLE 6.7 Denominator Polynomials for Transfer Functions of Bessel Filters Normalized to Have Unit Delay at $\omega = 0$

n	$q_n(s)$
2	$s^2 + 3s + 3$
3	$s^3 + 6s^2 + 15s + 15$
4	$s^4 + 10s^3 + 45s^2 + 105s + 105$
5	$s^5 + 15s^4 + 105s^3 + 420s^2 + 945s + 945$
6	$s^6 + 21s^5 + 210s^4 + 1260s^3 + 4725s^2 + 10{,}395s + 10{,}395$
7	$s^7 + 28s^6 + 378s^5 + 3150s^4 + 17{,}325s^3 + 62{,}370s^2 + 135{,}135s + 135{,}135$
8	$s^8 + 36s^7 + 630s^6 + 6930s^5 + 9450s^4 + 270{,}270s^3 + 945{,}945s^2 + 2{,}027{,}025s + 2{,}027{,}025$

Unlike the transfer function for Butterworth and Chebyshev filters, Eq. (6.15) does not provide an explicit expression for the poles of the Bessel filter. The numerator of (6.15) will be a polynomial in s, upon which numerical root-finding methods (such as Algorithm 5.1) must be used to determine the pole locations for $H(s)$. Table 6.8 lists approximate pole locations for $n = 2$ through $n = 8$.

The transfer functions given by (6.15) are for Bessel filters normalized to have unit delay at $\omega = 0$. The poles p_k and denominator coefficients b_k can be

TABLE 6.8 Poles of Bessel Filter Normalized to Have Unit Delay at $\omega = 0$

n	Pole values
2	$-1.5 \pm 0.8660j$
3	-2.3222 $-1.8390 \pm 1.7543j$
4	$-2.1039 \pm 2.6575j$ $-2.8961 \pm 0.8672j$
5	-3.6467 $-2.3247 \pm 3.5710j$ $-3.3520 \pm 1.7427j$
6	$-2.5158 \pm 4.4927j$ $-3.7357 \pm 2.6263j$ $-4.2484 \pm 0.8675j$
7	-4.9716 $-2.6857 \pm 5.4206j$ $-4.0701 \pm 3.5173j$ $-4.7584 \pm 1.7393j$
8	$-5.2049 \pm 2.6162j$ $-4.3683 \pm 4.4146j$ $-2.8388 \pm 6.3540j$ $-5.5878 \pm 0.8676j$

TABLE 6.9 Factors for Renormalizing Bessel Filter Poles from Unit Delay at $\omega = 0$ to 3-dB Attenuation at $\omega = 1$

n	A
2	1.35994
3	1.74993
4	2.13011
5	2.42003
6	2.69996
7	2.95000
8	3.17002

renormalized for a 3-dB frequency of $\omega = 1$ by using

$$p'_k = Ap_k \qquad b'_k = A^{n-k} b_k$$

where the value of A appropriate for n is selected from Table 6.9. [The values from the table have been incorporated into the **BesselCoefficient( )** function.]

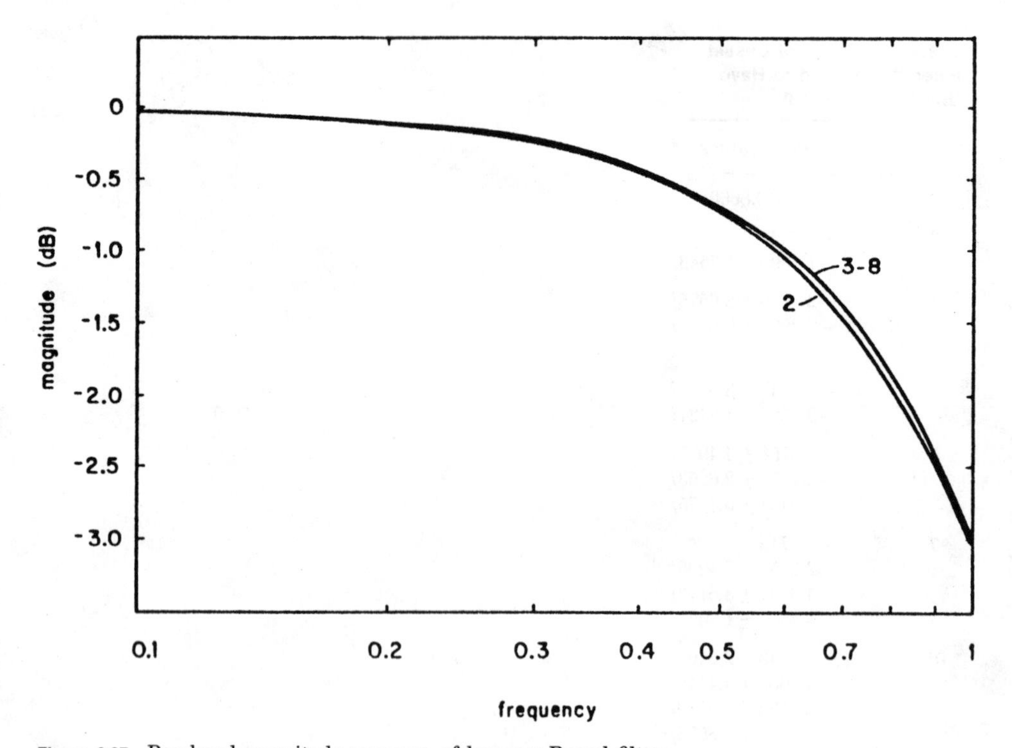

Figure 6.37 Passband magnitude response of lowpass Bessel filters.

Frequency response

Figures 6.37 and 6.38 show the magnitude responses for Bessel filters of several different orders. The frequency response data was generated by the C routine **BesselFreqResponse( )**, which is provided in Listing 6.7.

```
/**************************************/
/*                                    */
/*   Listing 6.7                      */
/*                                    */
/*   BesselFreqResponse()             */
/*                                    */
/**************************************/
#include <math.h>
#include "globDefs.h"
#include "protos.h"


int BesselFreqResponse(   int order,
                          real coef[],
                          real frequency,
                          real *magnitude,
                          real *phase)
{
struct complex numer, omega, denom, transferFunction;
int i;

numer = cmplx( coef[0], 0.0);
omega = cmplx( 0.0, frequency);
denom = cmplx( coef[order], 0.0);

for( i=order-1; i>=0; i--)
    {
    denom = cMult(omega,denom);
    denom.Re = denom.Re + coef[i];
    }
transferFunction = cDiv( numer, denom);

*magnitude = 20.0 * log10(cAbs(transferFunction));
*phase = 180.0 * arg(transferFunction) / PI;
return;
}
```

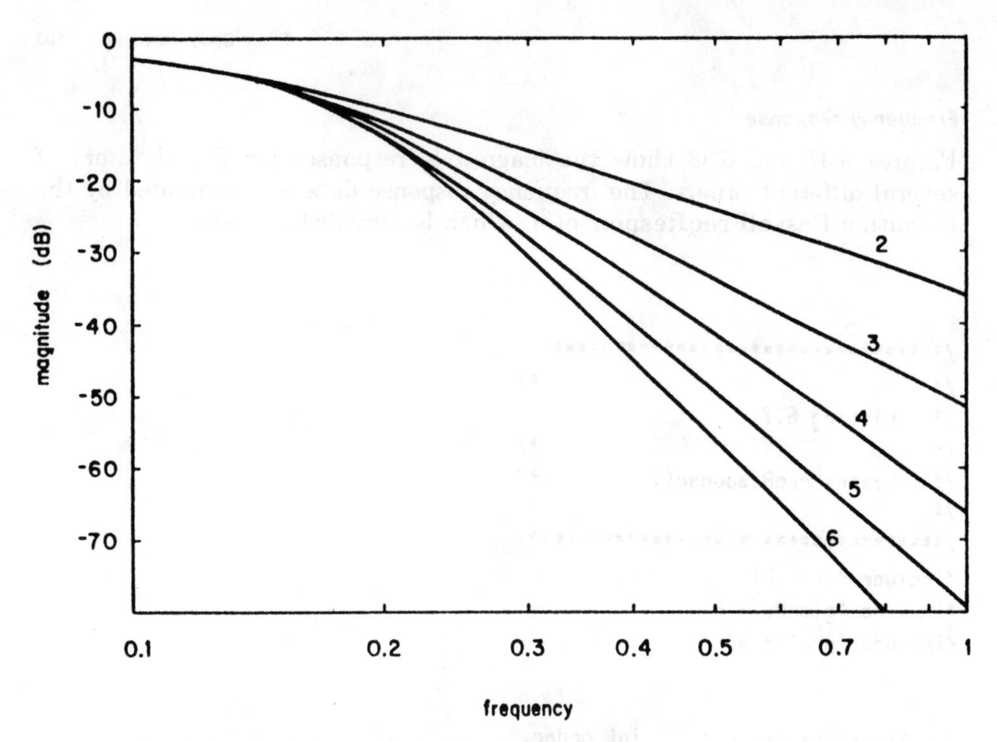

Figure 6.38 Stopband magnitude response of lowpass Bessel filters.

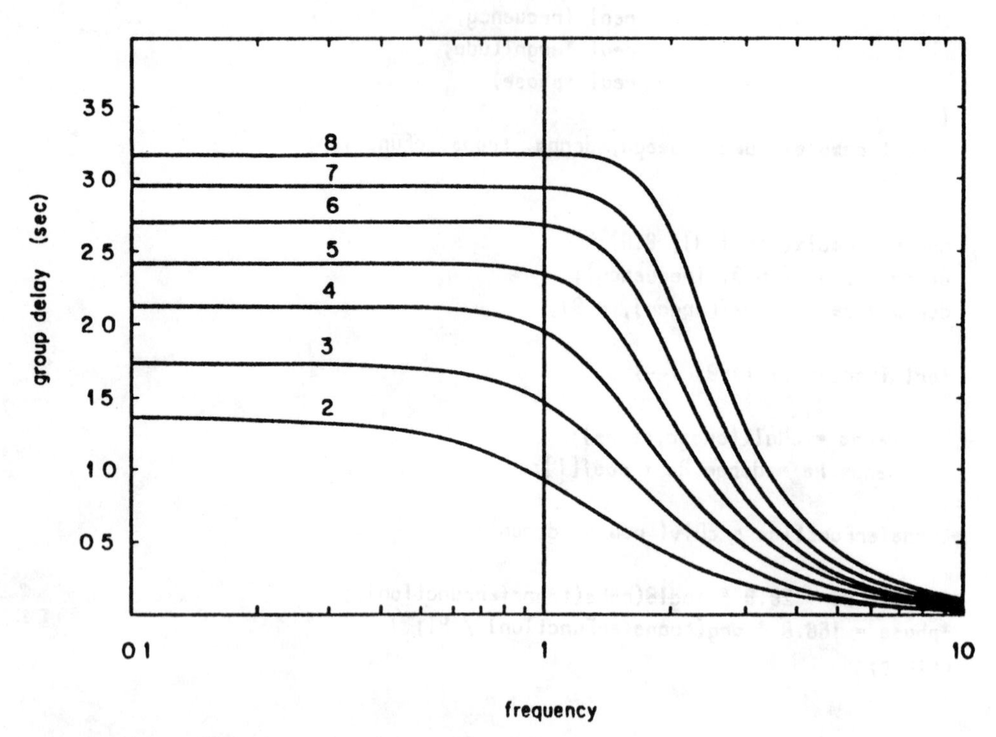

Figure 6.39 Group-delay response of lowpass Bessel filters.

Group delay

Group delays for lowpass Bessel filters of several different orders are plotted in Fig. 6.39. The data for these plots was generated by numerical differentiation of the phase response.

Chapter

7

Matrix Methods

Entire books have been written about matrices, matrix operations, and matrix properties; but for purposes of circuit analysis, we need to use only a relatively small subset of this material. After laying a foundation of notational practices, basic terminology, and fundamental operations, this chapter presents the matrix techniques that are commonly used for numerical solution of network equations. Cramer's rule is best suited for hand computations involving small matrices, while the remaining methods are more suited for computer implementation.

7.1 Matrix Fundamentals

A matrix is a rectangular array of *elements*, where each element can be either a numeric value or an algebraic expression. Individual "nonmatrix" values or expressions such as these elements are referred to as *scalars* whenever it is desired to emphasize their nonmatrix nature. It is also customary to denote individual elements of a matrix by subscripted lowercase letters—the element in row 2 and column 3 of matrix **A** is denoted as a_{23} or $a_{2,3}$. Commas are used in the subscript whenever necessary to avoid confusion. The usual practice is to denote general matrices by boldface uppercase roman letters. Scalar values are usually denoted by lowercase italic nonboldface roman letters. Shown below are five matrices **A**, **B**, **C**, **x**, and **z**:

$$\mathbf{A} = \begin{bmatrix} 0 & 5 & 2 \\ 1 & 4 & 3 \\ 2 & -1 & 12 \end{bmatrix} \qquad \mathbf{B} = \begin{bmatrix} b_{1,1} & b_{1,2} & b_{1,3} \\ b_{2,1} & b_{2,2} & b_{2,3} \\ b_{3,1} & b_{3,2} & b_{3,3} \end{bmatrix} \qquad \mathbf{x} = \begin{bmatrix} x_1 \\ x_2 \\ x_3 \end{bmatrix}$$

$$\mathbf{C} = \begin{bmatrix} 1+3j & 15-7j \\ -2j & 5 \end{bmatrix} \qquad \mathbf{z} = [z_1 \quad z_2 \quad z_3]$$

Each horizontal line of elements is called a *row*, and each vertical line of elements is called a *column*. Matrix **A** has 3 rows and 3 columns and is therefore said to be a 3 × 3 (read "3 by 3") matrix. A matrix such as **A** having an equal number of rows and columns is referred to as a *square* matrix. A matrix such as **x** having only one column can be more specifically identified as a *column vector*, and a matrix such as **z** having only one row can be more specifically identified as a *row vector*. Column vectors and row vectors are usually denoted by boldface *lowercase* roman letters.

Matrix operations

A number of mathematical operations are defined for matrices. In this book, we use *scalar multiplication*, *matrix addition*, and *matrix multiplication*.

Scalar multiplication. A matrix can have each of its elements multiplied by a scalar. For the matrices defined above we can obtain

$$2\mathbf{A} = \begin{bmatrix} 0 & 10 & 4 \\ 2 & 8 & 6 \\ 4 & -2 & 24 \end{bmatrix} \qquad s\mathbf{B} = \begin{bmatrix} sb_{1,1} & sb_{1,2} & sb_{1,3} \\ sb_{2,1} & sb_{2,2} & sb_{2,3} \\ sb_{3,1} & sb_{3,2} & sb_{3,3} \end{bmatrix}$$

Matrix addition. Matrices of like dimensions can be added on an element-by-element basis. Thus,

$$\mathbf{B} + \mathbf{A} = \mathbf{F}$$

where

$$\mathbf{F} = \begin{bmatrix} b_{1,1} & b_{1,2} + 5 & b_{1,3} + 2 \\ b_{2,1} + 1 & b_{2,2} + 4 & b_{2,3} + 3 \\ b_{3,1} + 2 & b_{3,2} - 1 & b_{3,3} + 12 \end{bmatrix}$$

Because **A** and **C** do not have the same dimensions, they cannot be added.

Matrix multiplication. One might guess that matrix multiplication would be defined in a way that is analogous to matrix addition, i.e., corresponding elements of two like-dimensioned matrices multiplied on an element-by-element basis. This approach might seem logical, but matrix multiplication is defined somewhat differently. For two matrices **P** and **Q**, the matrix product **PQ** can be obtained if and only if the number of columns in **P** equals the number of rows in **Q**. The matrix resulting from the multiplication will have the same number of rows as **P** and the same number of columns as **Q**. For **PQ** = **R**, each element in **R** is obtained as

$$r_{ij} = \sum_{k=1}^{m} p_{ik} q_{kj}$$

where m is the number of columns in **P**.

Example 7.1 For the specific matrices defined above, the product **AC** is not defined. The products **Ax**, **xz**, and **zx** are given by

$$\mathbf{Ax} = \begin{bmatrix} 5x_2 + 2x_3 \\ x_1 + 4x_2 + 3x_3 \\ 2x_1 - x_2 + 12x_3 \end{bmatrix} \qquad \mathbf{xz} = \begin{bmatrix} x_1 z_1 \\ x_2 z_2 \\ x_3 z_3 \end{bmatrix}$$

$$\mathbf{zx} = [z_1 x_1 + z_2 x_2 + z_3 x_3]$$

Note that matrix multiplication is *not* commutative!

Systems of linear equations. A linear system of equations can be expressed in matrix form. Consider the following system of equations:

$$\begin{aligned} a_{11}x_1 + a_{12}x_2 + \cdots + a_{1n}x_n &= b_1 \\ a_{21}x_1 + a_{22}x_2 + \cdots + a_{2n}x_n &= b_2 \\ \cdots\cdots\cdots\cdots\cdots\cdots\cdots\cdots \\ a_{n1}x_1 + a_{n2}x_2 + \cdots + a_{nn}x_n &= b_n \end{aligned} \tag{7.1}$$

This system of *scalar* equations can be expressed as a single *matrix* equation

$$\mathbf{Ax} = \mathbf{b} \tag{7.2}$$

where

$$\mathbf{A} = \begin{bmatrix} a_{11} & a_{12} & \cdots & a_{1n} \\ a_{21} & a_{22} & \cdots & a_{2n} \\ \cdots & \cdots & \cdots & \cdots \\ a_{n1} & a_{n2} & \cdots & a_{nn} \end{bmatrix} \qquad \mathbf{x} = \begin{bmatrix} x_1 \\ x_2 \\ \vdots \\ x_n \end{bmatrix} \qquad \mathbf{b} = \begin{bmatrix} b_1 \\ b_2 \\ \vdots \\ b_n \end{bmatrix}$$

The following four elementary operations can be performed on Eq. (7.2) without changing the solution elements:

1. Swap the positions of two rows in **A** while simultaneously swapping the corresponding elements in **b**.
2. Multiply a row in **A** and the corresponding element in **b** by a nonzero constant.
3. Add to one row of **A** a linear combination of any of the other rows while simultaneously adding the corresponding combination of elements to the corresponding element in **b**.
4. Swap the position of two columns in **A** while simultaneously swapping the corresponding elements in **x**.

7.2 Cramer's Rule

For a matrix equation of the form $\mathbf{Ax} = \mathbf{b}$, where $\mathbf{A}$ and $\mathbf{b}$ are known, Cramer's rule can be used to solve for one or more of the unknown entries in $\mathbf{x}$. Specifically, x_k is given by

$$x_k = \frac{\det|\mathbf{A}_k|}{\det|\mathbf{A}|}$$

where $\mathbf{A}_k$ denotes the matrix formed by substituting $\mathbf{b}$ for the kth column in $\mathbf{A}$. This method is useful for hand calculations involving small matrices, but obtaining the determinants for moderate- to large-size matrices proves to be computationally more expensive than some other methods that can be used to solve the original equation.

Example 7.2 Given the matrix equation

$$\begin{bmatrix} 4 & 2 & 0 \\ 7 & 3 & 1 \\ -5 & 1 & 1 \end{bmatrix} \begin{bmatrix} x_1 \\ x_2 \\ x_3 \end{bmatrix} = \begin{bmatrix} 2 \\ 0 \\ 5 \end{bmatrix}$$

use Cramer's rule to find x_1, x_2, and x_3.

solution

$$\det|\mathbf{A}| = \det \begin{bmatrix} 4 & 2 & 0 \\ 7 & 3 & 1 \\ -5 & 1 & 1 \end{bmatrix} = -16$$

$$\det|\mathbf{A}_1| = \det \begin{bmatrix} 2 & 2 & 0 \\ 0 & 3 & 1 \\ 5 & 1 & 1 \end{bmatrix} = 14 \qquad v_1 = \frac{\det|\mathbf{A}_1|}{\det|\mathbf{A}|} = \frac{-7}{8}$$

$$\det|\mathbf{A}_2| = \det \begin{bmatrix} 4 & 2 & 0 \\ 7 & 0 & 1 \\ -5 & 5 & 1 \end{bmatrix} = -44 \qquad v_2 = \frac{\det|\mathbf{A}_2|}{\det|\mathbf{A}|} = \frac{11}{4}$$

$$\det|\mathbf{A}_3| = \det \begin{bmatrix} 4 & 2 & 2 \\ 7 & 3 & 0 \\ -5 & 1 & 5 \end{bmatrix} = 34 \qquad v_3 = \frac{\det|\mathbf{A}_3|}{\det|\mathbf{A}|} = \frac{-17}{8}$$

7.3 Gaussian Elimination

Consider a matrix equation of the form

$$\mathbf{Ax} = \mathbf{b} \tag{7.3}$$

where $\mathbf{A}$ is an $n \times n$ matrix and $\mathbf{x}$ and $\mathbf{b}$ are n-element column vectors. Usually the elements of $\mathbf{A}$ and $\mathbf{b}$ are given, and we must solve for $\mathbf{x}$. Gaussian elimination is a strategy for the orderly, systematic application of operations 1, 2, and 3 from Sec. 7.1 to obtain the desired solution. The basic idea is most easily conveyed by means of a concrete example. Consider the following matrix equation:

$$\begin{bmatrix} 4 & 2 & 0 \\ 7 & 3 & 1 \\ -5 & 1 & 1 \end{bmatrix} \begin{bmatrix} x_1 \\ x_2 \\ x_3 \end{bmatrix} = \begin{bmatrix} 2 \\ 0 \\ 5 \end{bmatrix} \tag{7.4}$$

Begin by augmenting **A** with **b**:

$$\left[\begin{array}{ccc|c} 4 & 2 & 0 & 2 \\ 7 & 3 & 1 & 0 \\ -5 & 1 & 1 & 5 \end{array}\right] \tag{7.5}$$

Subtract $7/4$ times row 1 from row 2 to create a zero in column 1, row 2.

$$\left[\begin{array}{ccc|c} 4 & 2 & 0 & 2 \\ 0 & -1/2 & 1 & -7/2 \\ -5 & 1 & 1 & 5 \end{array}\right] \tag{7.6}$$

Create a zero in column 1 of row 3 by adding $5/4$ times row 1 to row 3.

$$\left[\begin{array}{ccc|c} 4 & 2 & 0 & 2 \\ 0 & -1/2 & 1 & -7/2 \\ 0 & 7/2 & 1 & 15/2 \end{array}\right] \tag{7.7}$$

Finally, create a zero in column 2, row 3 by adding 7 times row 2 to row 3.

$$\left[\begin{array}{ccc|c} 4 & 2 & 0 & 2 \\ 0 & -1/2 & 1 & -7/2 \\ 0 & 0 & 8 & -34/2 \end{array}\right] \tag{7.8}$$

This ends the *forward elimination* phase of the procedure. The **A** matrix is now in upper triangular form, and we begin *backward substitution* to obtain the elements of **x**. The bottom row in (7.8) corresponds to the scalar equation

$$8x_3 = -34/2 \qquad \text{or} \qquad x_3 = -17/8 \tag{7.9}$$

The second row corresponds to the scalar equation

$$-1/2\, x_2 + x_3 = -7/2 \tag{7.10}$$

Substituting (7.9) into (7.10), we find $x_2 = 11/4$. Finally we take the equation represented by the top row

$$4x_1 + 2x_2 = 2$$

and solve for $x_1 = {}^-7/_8$. Algorithm 7.1 details the gaussian elimination procedure for a system of n equations. Steps 1 through 3 comprise the forward elimination phase, and steps 4 and 5 comprise the backward substitution phase. This algorithm is labeled as "naive" since it has some weaknesses that will have to be corrected before it can be implemented as a working software function.

Algorithm 7.1 (Naive gaussian elimination). This algorithm finds the n-element solution vector $\mathbf{x}$ for a matrix equation $\mathbf{Ax} = \mathbf{b}$, where $\mathbf{A}$ is an $n \times n$ matrix with known elements and $\mathbf{b}$ is an n-element column vector with known elements.

Step 1. Augment the $\mathbf{A}$ matrix by appending the $\mathbf{b}$ vector on the right. Denote the augmented matrix as $\mathbf{C}$.

Step 2. Designate row 1 as the *working row* and column 1 as the *working column*. The intersection of the working row and working column is called the *working position*. Let i denote the index of the working position: $i \leftarrow 1$. (The matrix element c_{ii} located in the working position is called the *pivot*, or *pivot element*.)

Step 3. From each row k that lies below the working row (i.e., for $k = i + 1, i + 2, \ldots, n$), subtract s times row i where $s = c_{kj}/c_{ij}$. (This operation will place zeros in column j of rows $i + 1$ through n.)

Step 4. If $i > n - 1$, go to step 5; otherwise designate a new working position: $i \leftarrow i + 1$, and return to step 3.

Step 5. At this point, the modified submatrix $\mathbf{A}$ will be in upper triangular form. Set $x_n \leftarrow b_n / a_{nn}$, where $a_{nn} = c_{nn}$ and $b_n = c_{n,n+1}$.

Step 6. For $j = n - 1, n - 2, \ldots, 1$, set

$$x_j \leftarrow \frac{b_j - \sum_{k=j+1}^{n} a_{jk} x_k}{a_{jj}}$$

where $\quad a_{jk} = c_{jk} \qquad b_j = c_{j,n+1}$ ■

Practical gaussian elimination

The most blatant limitation of Algorithm 7.1 becomes evident when a pivot value of zero is encountered, since this would entail division by zero in step 3. There are two ways to deal with such an event:

- Swap the row that contains the offending pivot with an acceptable row from those below it.
- Swap the column that contains the offending pivot with an acceptable column from those to the right. The corresponding elements in $\mathbf{x}$ must also be swapped.

A pivot element need not be zero to cause difficulties—a different, somewhat more subtle problem can occur as a result of computer rounding when relatively small pivot values are encountered. Consider the following system of two equations:

$$\begin{aligned} 10^{-8}x_1 + x_2 &= 1 \\ x_1 + x_2 &= 2 \end{aligned} \tag{7.11}$$

Manual application of naive gaussian elimination yields

$$\left[\begin{array}{cc|c} 10^{-8} & 1 & 1 \\ 1 & 1 & 2 \end{array}\right]$$

$$\left[\begin{array}{cc|c} 10^{-8} & 1 & 1 \\ 0 & 1-10^8 & 2-10^8 \end{array}\right]$$

$$x_2 = \frac{2-10^8}{1-10^8} = \frac{-99{,}999{,}998}{-99{,}999{,}999} = .99999998\overline{99999998}\cdots \tag{7.12}$$

$$x_1 = 10^8(1-x_2) = 10^8\left(1 - \frac{2-10^8}{1-10^8}\right) = 1.00000001\overline{00000001}\cdots \tag{7.13}$$

For a computer implementation, the problem lies in the terms $2-10^8$ and $1-10^8$. In most typical floating-point numeric formats, there are only around 24 bits in the mantissa. In hexadecimal notation, our two terms of concern are

$$99{,}999{,}998 = \text{5F5E0FEh}$$

$$99{,}999{,}999 = \text{5F5E0FFh}$$

Assuming that our computer retains only the most significant 24 bits (six hexadecimal digits) of floating-point numbers, both the numerator and denominator of Eq. (7.12) will be truncated to 5F5E0F0 hexadecimal or 99,999,984 decimal. Therefore, x_2 will be evaluated as 1.00, and Eq. (7.13) will yield $x_1 = 0$, which is incorrect. Using a numeric format with more bits of precision is one solution to this problem, but it is not the only solution. Next we will see how a significant improvement can be obtained by making a small change in the gaussian elimination algorithm.

The sequence of the equations in system (7.11) is purely arbitrary—we could just as easily have written

$$\begin{aligned} x_1 + x_2 &= 2 \\ 10^{-8}x_1 + x_2 &= 1 \end{aligned}$$

For this permuted system, manual application of naive gaussian elimination yields

$$\left[\begin{array}{cc|c} 1 & 1 & 2 \\ 10^{-8} & 1 & 1 \end{array}\right]$$

$$\left[\begin{array}{cc|c} 1 & 1 & 2 \\ 0 & 1-10^{-8} & 1-2\times 10^{-8} \end{array}\right]$$

$$x_2 = \frac{1-2\times 10^{-8}}{1-10^{-8}} \tag{7.14}$$

Assuming a 24-bit mantissa, both the numerator and denominator of (7.14) will be equal to either 1 or $1-\epsilon$, depending upon the rounding or truncation approach built into the computer's floating-point implementation. In any event, they will be equal to each other, and x_2 will be evaluated as exactly 1.0. Then x_1 will be found as

$$x_1 + x_2 = 2 \Leftrightarrow x_1 = 2 - x_2 = 2 - 1 = 1$$

These values of x_1 and x_2 are "close enough" to the values of (7.12) and (7.13) for any circuit analysis or design problems encountered in the real world.

Based on the above, it appears that all we need to do is modify the algorithm for gaussian elimination to ensure that the rows are processed in the "best" order. Actually, determination of a "best" order is, in general, extremely difficult; but it is fairly easy to determine a processing sequence that is "good enough." In fact, there are three different *pivoting* strategies—partial pivoting, scaled partial pivoting, and full pivoting—which offer increasing performance at the expense of increasing complexity.

Partial pivoting

It is not possible to use a pivot of zero. Therefore, if the working row contains a zero in the working column (i.e., in the pivot position), as discussed above, we must exchange either the offending row or the offending column. This concept can be extended to avoid use of pivots that are nearly zero as well as those that equal zero. Rather than trying to decide how small is too small, the simplest approach is to take as the working row whichever of the remaining rows has the largest absolute value in the working column. This approach is called *partial pivoting*.

Example 7.3 Consider the augmented matrix (7.5) which is reproduced here for convenience:

$$\left[\begin{array}{ccc|c} 4 & 2 & 0 & 2 \\ 7 & 3 & 1 & 0 \\ -5 & 1 & 1 & 5 \end{array}\right]$$

In accordance with the partial pivoting strategy, we select 7 as the pivot and reorder the matrix as

$$\left[\begin{array}{rrr|r} 7 & 3 & 1 & 0 \\ 4 & 2 & 0 & 2 \\ -5 & 1 & 1 & 5 \end{array}\right]$$

We then subtract the appropriate multiple of row 1 from rows 2 and 3 to obtain

$$\left[\begin{array}{rrr|r} 7 & 3 & 1 & 0 \\ 0 & 2/7 & -4/7 & 2 \\ 0 & 22/7 & 12/7 & 5 \end{array}\right]$$

Now row 2 is the working row, but since $22/7 > 1/7$, we exchange rows 2 and 3.

$$\left[\begin{array}{rrr|r} 7 & 3 & 1 & 0 \\ 0 & 22/7 & 12/7 & 5 \\ 0 & 2/7 & -4/7 & 2 \end{array}\right]$$

Finally we subtract $1/22$ times row 2 from row 3 to obtain

$$\left[\begin{array}{rrr|r} 7 & 3 & 1 & 0 \\ 0 & 22/7 & 12/7 & 5 \\ 0 & 0 & -8/11 & 17/11 \end{array}\right]$$

Algorithm 7.2 (Forward elimination phase for gaussian elimination with partial pivoting). This algorithm details the forward elimination phase only—the backward substitution phase is identical to steps 5 and 6 of Algorithm 7.1.

Step 1. Augment the **A** matrix by appending the **b** vector on the right. Denote the augmented matrix as **C**.

Step 2. Designate row 1, column 1 as the *working position*. Let i denote the index of the working position: $i \leftarrow 1$.

Step 3. From rows i through n, find the row that has the largest magnitude in column i. Exchange this row with row i.

Step 4. If element $c_{ii} = 0$, the matrix is singular, and gaussian elimination will not be successful. If $c_{ii} \neq 0$, continue to step 5.

Step 5. From each row k that lies below the working row (i.e., for $k = i+1, i+2, \ldots, n$) subtract s times row i, where $s = c_{kj}/c_{ij}$. (This operation will place zeros in column j of rows $i+1$ through n.)

Step 6. If $i > n-1$, perform backward substitution; otherwise designate a new working position $i \leftarrow i+1$, and return to step 3. ■

Programming note 7.1

Algorithms 7.1, 7.2, and 7.3 call for the exchange of rows in the **C** matrix. Rather than actually moving the coefficients around in memory every time a row exchange is needed, it is far more efficient to use an array of indices to convert back and forth between "logical" row numbers and "physical" row numbers. Initially the physical row numbers and logical row numbers are equal, so we set **rowIndex[i]** = **i** for $i = 1, 2, \ldots, n$. Now suppose that one of the pivoting schemes calls for the exchange of rows 1 and 3. We simply execute

```
temp =rowIndex[3];
rowIndex[3]=rowIndex[1];
rowIndex[1]=temp;
```

This entails moving only three integers rather than the $3n$ floating-point values that would have to be moved to actually exchange the row elements. To access element 2 in new row 3, we write

```
C_matrix [ rowIndex [3] ] [2]
```

The extra level of indirection involved in this expression is not really significant, because in any operation that accesses each element in the row we would probably use a pointer:

```
/* This code fragment adds P times row 1 to row 2 */
   source_ptr=&( C_matrix [ rowIndex[1] ] [1]);
   dest_ptr=&( C_matrix [ rowIndex[2] ] [1]);
   for( i=1; i<=n; i++) *dest_ptr +=(P * (*source_ptr));
```

Scaled partial pivoting

The difficulties caused by small pivots actually depend upon how small the pivot is relative to the other coefficients in the same row. Therefore, an improved pivoting strategy called *scaled partial pivoting* first determines a scale factor for each row in the **A** matrix. This scale factor is equal to the coefficient having the largest magnitude. Then, when elements in the working column are being examined to determine the new pivot, each candidate element is first normalized by the scale factor for its respective row.

Example 7.4 Consider the following **C** matrix which has been further augmented on the right with a column containing the scale factor for each row.

$$\begin{bmatrix} 0.001 & 1.0 & 0 & \vdots & 1.001 & \vdots & 1.0 \\ 0 & 2.0 & 3.0 & \vdots & 4.0 & \vdots & 3.0 \\ 0.0001 & 0.0001 & 0 & \vdots & 0.0002 & \vdots & 0.0001 \end{bmatrix}$$

The ratios of the first element to the scale factor are 0.001, 0.0, and 1.0 for rows 1 through 3, respectively. Since 1.0 is the biggest of these, row 3 should be used as the pivot row. Exchanging row 1 with row 3 and then subtracting the appropriate multiples of new row 1 from row 2 and new row 3, we obtain

$$\begin{bmatrix} 0.0001 & 0.0001 & 0 & \vdots & 0.0002 & \vdots & 0.0001 \\ 0 & 2.0 & 3.0 & \vdots & 4.0 & \vdots & 3.0 \\ 0 & 0.999 & 0 & \vdots & 0.999 & \vdots & 1.0 \end{bmatrix}$$

Notice that the scale factors do not change—they move when their row moves, but they do not participate in the multiplying and subtracting. The ratios of the second element to the scale factor are 0.6667 and 0.999 for rows 2 and 3, so we must exchange rows 2 and 3. Using 0.999 as the pivot, we obtain

$$\left[\begin{array}{ccc|c|c} 0.0001 & 0.0001 & 0 & 0.0002 & 0.0001 \\ 0 & 0.999 & 0 & 0.999 & 1.0 \\ 0 & 0 & 3.0 & 2.0 & 3.0 \end{array}\right]$$

Algorithm 7.3 (Forward elimination phase for gaussian elimination with scaled partial pivoting)

Step 1. Augment the **A** matrix by appending the **b** vector on the right. Denote the augmented matrix as **C**.

Step 2. For each row k, find a scale factor s_k such that s_k equals the element in row k having the largest absolute value. If any of the s_k equals zero, the matrix is singular and gaussian elimination will not be successful.

Step 3. Designate row 1, column 1 as the working position. Let i denote the index of the working position: $i \leftarrow 1$.

Step 4. From rows i through n, find the row k for which $|c_{ki}/s_k|$ is the largest. Exchange this row with row i.

Step 5. From each row k that lies below the working row (i.e., for $k = i+1, i+2, \ldots, n$), subtract s times row i, where $s = c_{kj}/c_{ij}$.

Step 6. If $i \geq n-1$, perform backward substitution; otherwise designate a new working position $i \leftarrow i+1$, and return to step 4. ■

Full pivoting

In partial pivoting, the search for a new pivot considers all elements in the working column that lie in or below the working row. *Full pivoting* extends this search to include all elements in the working row that lie to the right of the working column. If the new pivot value is found below the working position, a row exchange operation must be performed. If the new pivot value is found to the right of the working position, a column exchange operation must be performed.

Example 7.5 Consider the following **C** matrix which has been annotated along the top to indicate the variable corresponding to each column:

$$\begin{array}{c} \begin{array}{ccc} x_1 & x_2 & x_3 \end{array} \\ \left[\begin{array}{ccc|c} 1.5 & 1.0 & 2.0 & 9.0 \\ -2.0 & 8.0 & 6.0 & -0.5 \\ -1.0 & 2.0 & 4.0 & -2.0 \end{array}\right] \end{array}$$

In the **A** submatrix, the element having the largest magnitude is 8.0. Performing row and column exchanges to bring this element into the pivot position yields

$$\begin{array}{c} \begin{array}{ccc} x_2 & x_1 & x_3 \end{array} \\ \left[\begin{array}{ccc|c} 8.0 & -2.0 & 6.0 & -0.5 \\ 1.0 & 1.5 & 2.0 & 9.0 \\ 2.0 & -1.0 & 4.0 & -2.0 \end{array}\right] \end{array}$$

We then subtract the appropriate multiples of row 1 from row 2 and row 3 to obtain

$$\begin{array}{ccc|c} x_2 & x_1 & x_3 & \\ \end{array}$$
$$\left[\begin{array}{ccc|c} 8.0 & -2.0 & 6.0 & -0.5 \\ 0.0 & 1.75 & 1.25 & 9.0625 \\ 0.0 & -0.5 & 2.5 & -1.875 \end{array}\right]$$

Of the remaining elements (i.e., those not in row 1 or column 1), the largest is 2.5 in row 3, column 3. Therefore, we exchange rows 2 and 3 and then exchange columns 2 and 3:

$$\begin{array}{ccc|c} x_2 & x_3 & x_1 & \\ \end{array}$$
$$\left[\begin{array}{ccc|c} 8.0 & 6.0 & -2.0 & -0.5 \\ 0.0 & 2.5 & -0.5 & -1.875 \\ 0.0 & 1.25 & 1.75 & 9.0625 \end{array}\right]$$

Finally, we subtract ½ times row 2 from row 3:

$$\begin{array}{ccc|c} x_2 & x_3 & x_1 & \\ \end{array}$$
$$\left[\begin{array}{ccc|c} 8.0 & 6.0 & -2.0 & -0.5 \\ 0.0 & 2.5 & -0.5 & -1.875 \\ 0.0 & 0.0 & 2.0 & 10.0 \end{array}\right]$$

$$x_1 = \frac{10.0}{2.0} = 5.0$$

$$x_3 = \frac{-1.875 - (-0.5)(5.0)}{2.5} = 0.25$$

$$x_2 = \frac{-0.5 - (-2.0)(5.0) - (6.0)(0.25)}{8.0} = 1.0$$

Algorithm 7.4 (Forward elimination phase for gaussian elimination with full pivoting)

Step 1. Augment the **A** matrix by appending the **b** vector on the right. Denote the augmented matrix as **C**.

Step 2. Designate row 1, column 1 as the working position. Let 1 denote the index of the working position: $i \leftarrow 1$.

Step 3. From elements c_{jk}, $j = i, i+1, \ldots, n$, $k = i, i+1, \ldots, n$, find row j and column k for which $|c_{ik}|$ is maximized.

Step 4. If $j \neq i$, exchange rows i and j.

Step 5. If $k \neq i$, exchange columns i and k and also exchange the corresponding elements x_j and x_k in **x**.

Step 6. From each row k that lies below the working row (i.e., for $k = i+1, i+2, \ldots, n$) subtract s times row i, where $s = c_{kj}/c_{ij}$.

Step 7. If $i \geq n-1$, perform backward substitution; otherwise designate a new working position $i \leftarrow i+1$, and return to step 4. ■

Comparison of pivoting strategies

Each of the pivoting strategies adds extra comparisons and/or divisions to the computational burden associated with solving a matrix equation via gaussian elimination. Partial pivoting adds $n(n-1)/2$ extra comparisons. Scaled partial pivoting adds $3n(n-1)/2$ extra comparisons and $n(n+1)/2-1$ extra divisions. Full pivoting adds $n(n-1)(2n+5)/6$ extra comparisons. Table 7.1 lists some specific number of additional operations for real matrices. For the case of complex-valued matrices we should bear in mind that complex operations involve significantly more computation than the corresponding real operations. A complex division entails six real multiplications, three real additions, and two real divisions. A complex magnitude operation involves two real multiplications, two real additions, and maybe one real square root operation. (See Programming note 7.2 for an explanation of the *maybe*.) A complex multiplication includes four real multiplications and two real additions. Table 7.2 lists the number of additional *real* operations that are required for pivoting in various sizes of *complex* matrices. Scaled partial pivoting is the most popular compromise between performance and economy. Listing 7.1 contains a **C** function that implements gaussian elimination with scaled partial pivoting.

```
/*****************************************************/
/*                                                   */
/*  Listing 7.1                                      */
/*                                                   */
/*  GaussianElimination()                            */
/*                                                   */
/*****************************************************/
#include "globDefs.h"
#include "protos.h"

int GaussianElimination( int n,
                         struct complex A[][MAX_COLUMNS],
                         struct complex b[],
                         struct complex x[])
{
int i, j, k, returnCode, bigIndex;
int index[MAX_COLUMNS];
struct complex z;
real y, big, s[MAX_COLUMNS];

returnCode = 0;
/*--------------------------------------*/
/*  Find scale factor for each row      */
```

```
for( i=1; i<=n; i++)
   {
   index[i]=i;
   big = 0.0;
   for( j=1; j<=n; j++)
      {
      if( cAbs( A[i][j] ) > big) big = cAbs(A[i][j]);
      }
   s[i] = big;
   if( big == 0.0 )
      {
      returnCode = 1;
      return returnCode;
      }
   }
for(i=1; i<n; i++)
   {
   big = 0.0;
   for( k=i; k<=n; k++)
      {
      y = cAbs(A[index[k]][i])/ s[k];
      if( y>big )
         {
         big = y;
         bigIndex = k;
         }
      }
   k = index[bigIndex];
   index[bigIndex] = index[i];
   index[i] = k;
   /*-------------------------------------------------*/
   /*  Subtract working row from all rows beneath it  */
   for(k=i+1; k<=n; k++)
      {
      z = cDiv(A[index[k]][i],A[index[i]][i]);
      A[index[k]][i] = cmplx(0.0,0.0);
      for(j=i+1; j<=n; j++)
         {
         A[index[k]][j] = cSub( A[index[k]][j],
                         cMult(z,A[index[i]][j]) );
         }
      b[index[k]]= cSub( b[index[k]], cMult(z,b[index[i]]) );
      }
   }
/*  Perform backwards substitution to solve for x[]  */
```

```
for( i=n; i>0; i--)
    {
    k=index[i];
    x[i] = b[k];
    for(j=i+1; j<=n; j++) x[i] = cSub( x[i],cMult(A[k][j], x[j]));
    x[i] = cDiv(x[i], A[k][i]);
    }
return returnCode;
}
```

TABLE 7.1 Additional Operations Required for Pivoting in Real-Valued Matrices

n	Comparisons for partial pivoting	Comparisons for scaled partial pivoting	Divisions for scaled partial pivoting	Comparisons for full pivoting
2	1	3	2	3
5	10	30	14	50
10	45	135	54	375
20	190	570	209	2,850
50	1,225	3,675	1,274	42,875
100	4,950	14,850	5,049	338,250

Programming note 7.2

The various pivoting strategies call for comparing the magnitudes of matrix elements. In circuit analysis problems, the matrix elements will generally be complex. Thus, the comparisons entail computation of magnitudes in accordance with the following formula:

$$|z| = \sqrt{(\text{Re } z)^2 + (\text{Im } z)^2} \tag{7.15}$$

Comparisons of real values are "no sweat," but calculation of (7.15) makes complex comparisons more costly. However, there is some relief; if $|z_1| > |z_2|$, then $|z_1|^2 > |z_2|^2$. Therefore, we can eliminate the square root operation and compare squared magnitudes. There are methods for computing $|z|$ that do not involve a square root operation, but since most of these involve two real multiplications and one real addition, *plus* one real comparison, we will stick with using $|z|^2$, which requires only two real multiplications plus one real addition.

7.4 Triangular Factorization

The *triangular factorization* or *LU factorization* technique for solving matrix equations is closely related to gaussian elimination but has certain

TABLE 7.2 Additional Real Operations Required for Pivoting in Complex-Valued Matrices

Partial Pivoting

n	Additions	Multiplications	Total
2	1	2	3
5	10	20	30
10	45	90	135
20	190	380	570
50	1,225	2,450	3,675
100	4,950	9,900	14,850

Scaled Partial Pivoting

n	Additions	Multiplications	Divisions	Total
2	9	18	4	31
5	72	144	28	244
10	296	592	108	996
20	1,197	2,394	418	4,009
50	7,497	14,994	2,548	25,039
100	29,997	59,994	10,098	100,089

Full Pivoting

n	Additions	Multiplications	Total
2	3	6	9
5	50	100	150
10	375	750	1,125
20	2,850	5,700	8,550
50	42,875	85,750	128,625
100	338,250	676,500	1,014,750

advantages that will prove useful in certain problems. Given the matrix equation

$$\mathbf{Ax} = \mathbf{b}$$

LU factorization assumes that the **A** matrix can be factored into two matrices **L** and **U** such that

$$\mathbf{LU} = \mathbf{A}$$

where **L** has any nonzero elements only on or below the main diagonal (that is, **L** is lower triangular) and **U** has any nonzero elements only on or above the main diagonal. In Crout's method all the diagonal elements of **U** are forced to equal 1, and in Doolittle's method all the diagonal elements of **L** are forced to be 1. The Choleski method requires that the diagonal of **L** equal the diagonal of **U**.

Algorithm 7.5 (Crout's method for triangular factorization). This algorithm factors a square matrix **A** into two matrices **L** and **U** such that **LU** = **A** and the diagonal elements of **L** equal 1.

Step 1. Set $l_{1,1} \leftarrow a_{1,1}$.
Step 2. For $j = 1, 2, \ldots, n$, set $u_{1,j} \leftarrow a_{1,j}/a_{1,1}$.
Step 3. For $i = 2, \ldots, n$, set $l_{i,1} \leftarrow a_{i,1}$.
Step 4. Set $k \leftarrow 2$.
Step 5. For $i = k, k+1, \ldots, n$, compute l_{ik}, using

$$l_{ik} = a_{ik} - \sum_{m=1}^{k-1} l_{im} u_{mk}$$

Step 6. Set $u_{kk} = 1$.
Step 7. If $k = n$, *stop*—the factorization is complete; otherwise continue to step 8.
Step 8. For $j = k+1, k+2, \ldots, n$, compute u_{kj}, using

$$u_{kj} = \frac{a_{kj} - \sum_{m=1}^{k-1} l_{km} u_{mj}}{l_{kk}}$$

Step 9. Set $k \leftarrow k+1$ and return to step 5. ■

Once the **A** matrix has been factored by using Algorithm 7.6, we have a matrix equation of the form

$$\mathbf{LUx} = \mathbf{b} \tag{7.16}$$

How do we go about solving for **x**? Let us define a new vector **z** such that

$$\mathbf{Ux} = \mathbf{z} \tag{7.17}$$

Then (7.16) can be written as

$$\mathbf{Lz} = \mathbf{b} \tag{7.18}$$

Since **L** is triangular, solving (7.18) for **z** is straightforward. The solution process begins at row 1 and is therefore called forward substitution. Once **z** is obtained, we perform backward substitution on (7.17) to solve for **x**.

Example 7.6 Using Algorithm 7.5, solve

$$\begin{bmatrix} 3/2 & 1 & 2 \\ -2 & 8 & 6 \\ -1 & 2 & 4 \end{bmatrix} \begin{bmatrix} x_1 \\ x_2 \\ x_3 \end{bmatrix} = \begin{bmatrix} 9 \\ -1/2 \\ -2 \end{bmatrix}$$

Step 1

$$l_{1,1} \leftarrow 3/2$$

Step 2

$$u_{1,1} \leftarrow 1, \quad u_{1,2} \leftarrow 2/3, \quad u_{1,3} \leftarrow 4/3$$

Step 3

$$l_{1,1} \leftarrow 3/2, \quad l_{1,2} \leftarrow -2, \quad l_{1,3} \leftarrow -1$$

Step 4

$$k \leftarrow 2$$

Step 5

$$\begin{aligned} l_{2,2} &\leftarrow a_{2,2} - l_{2,1} u_{1,2} \\ &= 8 - (-2)(2/3) = 28/3 \end{aligned}$$

$$\begin{aligned} l_{3,2} &\leftarrow a_{3,2} - l_{3,1} u_{1,3} \\ &= 2 - (-1)(2/3) = 8/3 \end{aligned}$$

Step 6

$$u_{2,2} \leftarrow 1$$

Step 7. Continue.
Step 8

$$\begin{aligned} u_{2,3} &= \frac{a_{2,3} - l_{2,1} u_{1,2}}{l_{2,2}} \\ &= \frac{6 - (-2)(4/3)}{28/3} = \frac{13}{14} \end{aligned}$$

Step 9

$$k \leftarrow 3$$

Step 5

$$\begin{aligned} l_{3,3} &\leftarrow a_{3,3} - l_{3,1} u_{1,3} - l_{3,2} u_{2,3} \\ &= 4 - (-1)(4/3) - (8/3)(13/14) = 20/7 \end{aligned}$$

Step 6

$$u_{3,3} \leftarrow 1$$

Step 7. Stop.

Collecting all the l and u elements, we have

$$\mathbf{L} = \begin{bmatrix} 3/2 & 0 & 0 \\ -2 & 28/3 & 0 \\ -1 & 8/3 & 20/7 \end{bmatrix} \qquad \mathbf{U} = \begin{bmatrix} 1 & 2/3 & 4/3 \\ 0 & 1 & 13/14 \\ 0 & 0 & 1 \end{bmatrix}$$

Foward substitution yields

$$\mathbf{Lz} = \mathbf{b}$$

$$z_1 = 9(2/3) = 6$$

$$z_2 = [-1/2 - (-2)(6)](3/28) = 69/56$$

$$z_3 = [-2 - (-1)(6) - (8/3)(69/56)](7/20) = 1/4$$

Backward substitution gives

$$\mathbf{Ux} = \mathbf{z}$$

$$x_3 = 1/4$$

$$x_2 = 69/56 - (13/14)(1/4) = 1$$

$$x_1 = 6 - (4/3)(1/4) - (2/3)(1) = 5$$

Because of the division by l_{kk} in step 8, Algorithm 7.5 will fail if l_{kk} equals zero for any k. It is possible to add pivoting to the basic algorithm so that a zero l_{kk} can be avoided for nonsingular matrices. While we're at it, we should design the pivoting strategy so as to minimize the effects of rounding errors.

Algorithm 7.6 (Doolittle's method with partial pivoting and "in-place" computation). This algorithm factors a square matrix **A** into two matrices **L** and **U** such that **LU** = **A** and the diagonal elements of **L** equal 1. The results are returned in **A**.

Step 1. Set $j = 1$.

Step 2. For $i = 1, 2, \ldots, j-1$, compute new a_{ij} (for **U**), using

$$a_{ij} = a_{ij} - \sum_{k=1}^{i-1} a_{ik} a_{kj}$$

Step 3. For $i = j, j+1, \ldots, N$, compute new a_{ij} (for **L**), using

$$a_{ij} = a_{ij} - \sum_{k=1}^{j-1} a_{ik} a_{kj}$$

Step 4. Set $i_{\max}$ equal to the value of i which maximizes

$$|a_{ij}| \quad \text{for } i \in \{j, j+1, \ldots, N\}$$

Step 5. If $i_{\max} \neq j$, then exchange row $i_{\max}$ and row j of **A**.

Step 6. If $j < N$, set $a_{ij} = a_{ij}/a_{jj}$ for $i = j+1, j+2, \ldots, N$.

Step 7. Set $d_j = i_{\max}$. (*Note:* d_j is not used by this algorithm, but it will be used by the forward-backward substitution procedures.)

Step 8. Set $j = j+1$; if new j is not greater than N, go to step 1; otherwise stop. ■

A C routine that implements Algorithm 7.6 is given in Listing 7.2, and a routine for forward-backward substitution is given in Listing 7.3.

```
/*****************************************************/
/*                                                   */
/*  Listing 7.2                                      */
/*                                                   */
/*  DoolittleMethod()                                */
/*                                                   */
/*****************************************************/

#include "globDefs.h"
#include "protos.h"

int DoolittleMethod( int N,
                     struct complex A[][MAX_COLUMNS],
                     int d[],
                     struct complex b[])
{
int i, j, k, iMax, returnCode;
struct complex cval;
real rval,maxA;
returnCode = 0;


/*  for loop corresponds to Steps 1 and 8 */
for( j=1; j<=N; j++)
    {

    /*---------- Step 2 follows ----------------*/
    for( i=1; i<j; i++)
        {
        cval = cmplx(0.0,0.0);
        for( k=1; k<i; k++) cval = cAdd(cval,cMult( A[i][k],A[k][j]));
        A[i][j] = cSub( A[i][j], cval );
        }
/*---------- Step 3 follows ---------------*/
for( i=j; i<=N; i++)
    {
    cval = cmplx(0.0,0.0);
    for( k=1; k<j; k++) cval = cAdd(cval,cMult( A[i][k],A[k][j]));
    A[i][j] = cSub( A[i][j], cval);
    }
```

```
/*---------- Step 4 follows ---------------*/
maxA = 0.0;
for( i=j; i<=N; i++)
    {
    rval = cAbs(A[i][j]);
    if( rval > maxA)
        {
        maxA = rval;
        iMax = i;
        }
    }
/*---------- Step 5 follows ---------------*/
if( iMax != j)
    {
    for( k=0; k<MAX_COLUMNS; k++)
        {
        cval = A[j][k];
        A[j][k] = A[iMax][k];
        A[iMax][k] = cval;
        }
    cval = b[j];
    b[j] = b[iMax];
    b[iMax] = cval;
    }
/*----------- Step 6 follows --------------*/
if( j<N )
    {
    for(i=j+1; i<=N; i++)
        {
        A[i][j] = cDiv( A[i][j], A[j][j]);
        }
        }
    /*------------ Step 7 follows -------------*/
    d[j] = iMax;
    }
return returnCode;
}

/**********************************************/
/*                                            */
/*   Listing 7.3                              */
/*                                            */
/*   SubstituteLU()                           */
/*                                            */
/**********************************************/
```

```
#include "globDefs.h"
#include "protos.h"
extern FILE *outFile;

int SubstituteLU( int n,
                    struct complex b[],
                    struct complex LU[][MAX_COLUMNS],
                    struct complex x[],
                    int d[])
{
int i, j, k, m, returnCode;
struct complex z[MAX_COLUMNS];
struct complex cval,cwork;

returnCode = 0;

for( i=1; i<=n; i++) /*  forward substitution using L matrix  */
    {
    cval = b[i];
    for( k=i-1; k>0; k--)
        {
        cval = cSub(cval, cMult(LU[i][k],z[k]));
        }
    z[i] = cval;
    }
for( i=n; i>=1; i--) /*  backward substitution using U matrix  */
    {
    cval = z[i];
    for( k=i+1; k<=n; k++)
        {
        cval = cSub(cval,cMult(LU[i][k],x[k]));
        }
    x[i] = cDiv( cval, LU[i][i]);
    }
return returnCode;
}
```

Chapter

8

Network Topology

Graph theory provides a precise means for describing network topology. This chapter presents those elements of graph theory that are involved in automatic generation of network equations as well as those elements that are useful in the derivation of computational analysis techniques.

8.1 Graphs

Consider the circuit shown in Fig. 8.1. The *oriented graph* or *directed graph* for this circuit is obtained by replacing each two-terminal element with a directed line segment, called an *edge*. As shown in Fig. 8.2, arrowheads are used to depict these edges. The direction of the arrowhead corresponds to the assumed direction of the current in the edge. For current sources, the direction is part of the source's specification. For voltage sources, the current is usually assumed to flow from the positive terminal to the negative terminal. For passive elements, the assumed direction of the current can be arbitrary. If an assumption is correct, the corresponding calculated value of current will turn out to be positive. Conversely, if the assumed direction is incorrect, the corresponding calculated current will be negative.

If all the arrowheads are removed from a directed graph, the remaining set of edges and nodes constitutes an *undirected* or *nonoriented* graph. A portion of a graph is called a *subgraph*. We often denote directed graphs, undirected graphs, and subgraphs as G_d, G_n, and G_s, respectively.

A *path* between nodes N_j and N_k in an undirected graph G_n can be loosely described as a direct route (via edges) from N_j to N_k. This route may contain multiple edges, but it is direct in the sense that it never loops back on itself. Formally, a path between node N_j and N_k in graph G is defined as an ordered set of edges $e_1, e_2, \ldots, e_n$ within G that satisfy the following conditions:

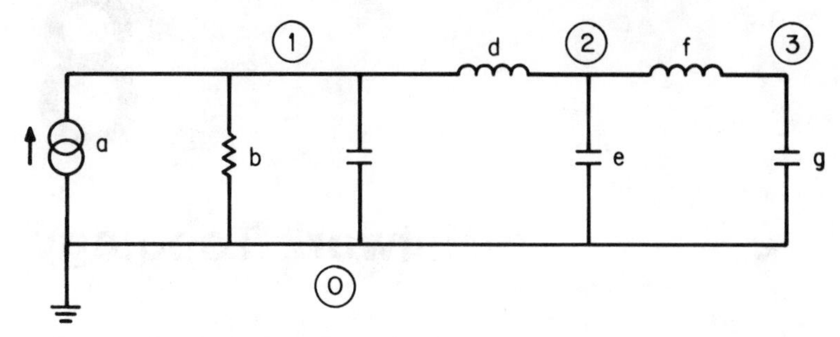

Figure 8.1 Circuit for discussion of undirected and directed graphs.

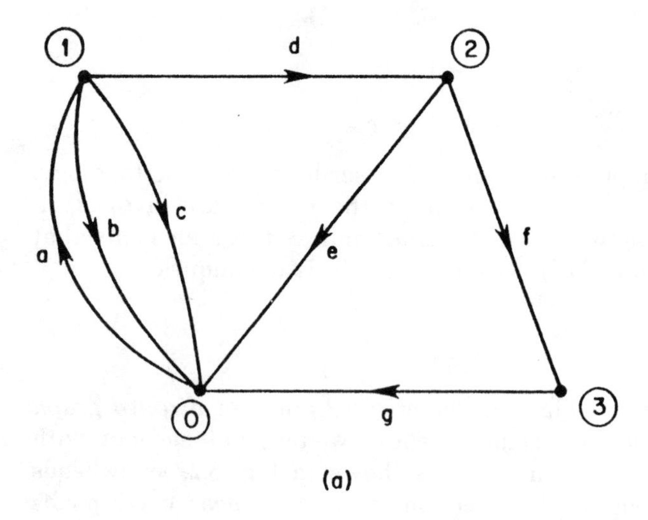

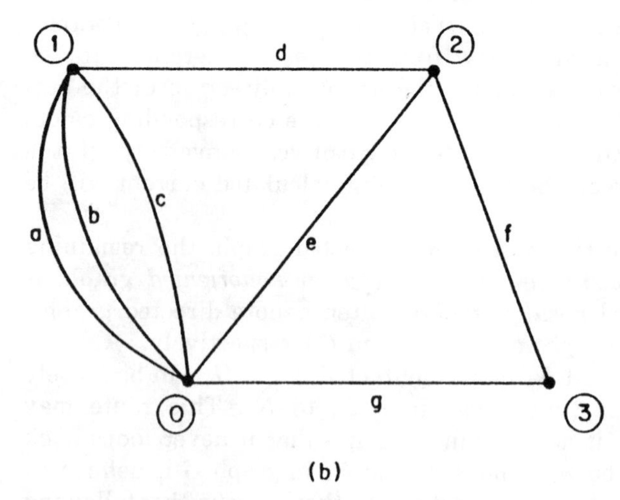

Figure 8.2 (*a*) Directed graph and (*b*) undirected graph for circuit of Fig. 8.1.

1. Consecutive edges e_i and e_{i+1} meet at a common node.
2. No more than two edges in the set meet at any given node.
3. Node N_j is the endpoint of one and only one edge.
4. Node N_k is the endpoint of one and only one edge.

Consider the undirected graph shown in Fig. 8.2*b*. The eight possible paths between node 1 and node 3 are $(d\,f)$, $(d\,e\,g)$, $(c\,e\,f)$, $(c\,g)$, $(b\,e\,f)$, $(b\,g)$, $(a\,e\,f)$, and $(a\,g)$. The edges $(a\,d\,f)$ do not form a path between nodes 1 and 3 because edge a starts at node 0. Edges $(a\,c\,d\,f)$ do not form a path because $(a\,c)$ constitutes a closed loop with edges b, c, and d all meeting at node 1, in violation of formal condition 2.

Loops and trees

An undirected graph G_n is described as *connected* if a path exists between every pair of nodes in the graph. A directed graph G_d is described as connected if the associated graph G_n is connected. A *loop* is any subgraph that is connected and in which exactly two edges meet at each node. A *tree* is any subgraph G_s of an undirected graph G_n for which the following are true:

1. G_s is connected.
2. G_s contains no loops.
3. G_s contains all the nodes of G_n.

The edges in the tree are called *tree branches* or *twigs*. The edges not belonging to the tree are called *links* or *chords*, and the set of all links forms a *cotree*. For a connected graph G_n having n nodes, any tree will have exactly $n-1$ twigs. In fact, *any* $n-1$ edges selected from G_n will constitute a tree provided that the edges are selected such that no loops are formed. For the graph shown in Fig. 8.2*b*, the edges $(c\,e\,g)$ form a tree with the corresponding cotree comprising edges $(a\,b\,d\,f)$. Edges $(c\,e)$ do not form a tree because they do not connect node 3. Edges $(a\,b\,d\,f)$ do not form a tree because edges $(a\,b)$ form a loop. Some other possible trees include $(a\,d\,f)$, $(d\,e\,f)$, and $(g\,a\,d)$.

Cutsets

A set of edges from a connected graph G_n is a *cutset* of the graph provided that

1. Removing the set of edges from G_n leaves a graph which is not connected.
2. Removing any portion of the set other than the entire set leaves a graph which remains connected.

For the graph shown in Fig. 8.2*b*, $(f\,g)$ is a cutset as are $(a\,b\,c\,d)$, $(d\,e\,f)$, and $(d\,e\,g)$. The edges $(d\,e\,f\,g)$ do not form a cutset—the removal of these edges creates an unconnected graph, but the removal of any three of the four edges also creates an unconnected graph.

8.2 Incidence Matrix

The connectivity information contained in a directed graph can be represented in matrix form as an *incidence matrix*. For a directed graph having n nodes and b branches, the incidence matrix will be an $n \times b$ matrix with a_{ij}, the element in row i, column j, given by

$$a_{ij} = \begin{cases} 1 & \text{if branch } j \text{ connects to, and is directed away from, node } i \\ -1 & \text{if branch } j \text{ connects to, and is directed toward, node } i \\ 0 & \text{if branch } j \text{ does not connect to node } i \end{cases}$$

Example 8.1 Construct the incidence matrix for the directed graph shown in Fig. 8.2.

solution There are four nodes, numbered 0 through 3, so the incidence matrix will have four rows. There are seven branches, labeled a through g, so there will be seven columns in the incidence matrix. If branches are assigned to columns in alphabetical order, the incidence matrix is

$$\mathbf{A} = \begin{bmatrix} 1 & -1 & -1 & 0 & -1 & 0 & -1 \\ -1 & 1 & 1 & 1 & 0 & 0 & 0 \\ 0 & 0 & 0 & -1 & 1 & 1 & 0 \\ 0 & 0 & 0 & 0 & 0 & -1 & 1 \end{bmatrix}$$

Each branch of the directed graph begins at exactly one node and ends at exactly one node, so each column of the incidence matrix will have exactly one entry equal to 1 and exactly one entry equal to -1. The sum of entries in any column will always equal zero. Therefore, any single row can be deleted without loss of information, since the zero-sum property can be applied to each column to regenerate the elements of the missing row. An incidence matrix with one of its rows deleted is called a *reduced incidence matrix*.

8.3 Loop Matrix

The loops within a directed graph can be compactly represented in the form of a *loop matrix*.

Algorithm 8.1 (Construction of a loop matrix for a directed graph $\mathbf{G}_d$)

Step 1. Temporarily ignore the direction of the edges, thus treating $\mathbf{G}_d$ as though it were an undirected graph. Identify all the loops in this *undirected* graph, and assign each loop a direction. These loops will be called *oriented loops*.

Step 2. For a graph having b branches and n oriented loops, the loop matrix will be an $n \times b$ matrix **B** with entries b_{ij}.

Step 3. For each branch j and oriented loop i, set $b_{ij} = 1$ if branch j is in loop i and the directions of the loop and branch agree.

Step 4. For each branch j and oriented loop i, set $b_{ij} = -1$ if branch j is in loop i and the directions of the loop and branch oppose.

Step 5. For each branch j and oriented loop i, set $b_{ij} = 0$ if branch j is not in loop i. ■

Example 8.2 Construct the loop matrix for the directed graph shown in Fig. 8.2.

solution Neglecting the direction of the edges, the following loops are identified:

$$L_1 = ab \quad L_2 = bc \quad L_3 = cde \quad L_4 = efg \quad L_5 = ac$$

$$L_6 = bde \quad L_7 = cdfg \quad L_8 = ade \quad L_9 = bdfg \quad L_{10} = adfg$$

The orientation of each loop is assumed to be clockwise. The loop matrix is then given by

$$\mathbf{B} = \begin{bmatrix} 1 & 1 & 0 & 0 & 0 & 0 & 0 \\ 0 & -1 & 1 & 0 & 0 & 0 & 0 \\ 0 & 0 & -1 & 1 & 1 & 0 & 0 \\ 0 & 0 & 0 & 0 & -1 & 1 & 1 \\ 1 & 0 & 1 & 0 & 0 & 0 & 0 \\ 0 & -1 & 0 & 1 & 1 & 0 & 0 \\ 0 & 0 & -1 & 1 & 0 & 1 & 1 \\ 1 & 0 & 0 & 1 & 1 & 0 & 0 \\ 0 & -1 & 0 & 1 & 0 & 1 & 1 \\ 1 & 0 & 0 & 1 & 0 & 1 & 1 \end{bmatrix}$$

8.4 Cutset Matrix

The cutsets within a directed graph can be compactly represented in the form of a *cutset matrix*.

Algorithm 8.2 **(Construction of a cutset matrix for a directed graph G_d)**

Step 1. Identify all the cutsets within the graph.

Step 2. Each cutset divides the graph into two disjoint pieces. Orient each cutset by assigning a direction relative to the two remaining pieces.

Step 3. For a graph having n cutsets and b branches, the cutset matrix will be an $n \times b$ matrix **D** having entries d_{ij}.

Step 4. For each cutset i and branch j, set $d_{ij} = 1$ if branch j is in cutset i and the directions of the branch and cutset agree.

Step 5. For each cutset i and branch j, set $d_{ij} = -1$ if branch j is in cutset i and the directions of the branch and cutset oppose.

Step 6. For each cutset i and branch j, set $d_{ij} = 0$ if branch j is not in cutset i. ■

Example 8.3 Construct the cutset matrix for the directed graph shown in Fig. 8.2.

solution The following cutsets are identified and assigned orientations as indicated:

Cutset 1: *abcd*, away from node 1
Cutset 2: *abceg*, away from node 0
Cutset 3: *def*, away from node 2
Cutset 4: *fg*, away from node 3
Cutset 5: *deg*, left to right
Cutset 6: *abcef*, top to bottom

The cutset matrix is then given by

$$\mathbf{D} = \begin{bmatrix} -1 & 1 & 1 & 1 & 0 & 0 & 0 \\ 1 & -1 & -1 & 0 & -1 & 0 & -1 \\ 0 & 0 & 0 & 1 & 1 & 1 & 0 \\ 0 & 0 & 0 & 0 & 0 & -1 & 1 \\ 0 & 0 & 0 & 1 & -1 & 0 & -1 \\ -1 & 1 & 1 & 0 & 1 & 1 & 0 \end{bmatrix}$$

Chapter

9

Network Equations

This chapter introduces several techniques for generating a set of equations which completely characterizes any circuit made up of resistors, capacitors, inductors, independent voltage sources, and independent current sources. These equations are then put into a matrix form that is convenient for solution via the techniques presented in Chap. 7. The algebraic approach involves using the Kirchhoff current law to write a current equation for each node in the circuit and then performing algebraic manipulations to put the set of equations into matrix form. A second approach involves a set of rules which allows the matrix equation to be written directly by inspection of the circuit to be analyzed. These rules lead directly to an algorithm suitable for computer generation of the network equations. A C program which implements this algorithm is then presented. The nodal admittance formulation is easy to implement, but unfortunately it only works for a limited set of component types. In Sec. 9.3, we examine an extension of the nodal admittance approach in which the admittance matrix is augmented with additional information for component types not having an admittance description.

9.1 Nodal Admittance Formulation

The *Kirchhoff current law* (KCL) states that the sum of currents entering a node equals the sum of currents leaving the node. This fact allows us to write an equation for each node in a circuit by setting the sum of currents entering the node from independent current sources equal to the sum of currents leaving the node via resistors, capacitors, and inductors. If we limit ourselves to consideration of just steady-state analyses, the equations can be formulated directly in the complex-frequency domain.

Consider the circuit fragment shown in Fig. 9.1. The current entering node 1 from the current source is I. Let V_1, V_2, V_3, and V_4 denote the voltages at nodes 1, 2, 3, and 4, respectively. (The values of these voltages are usually

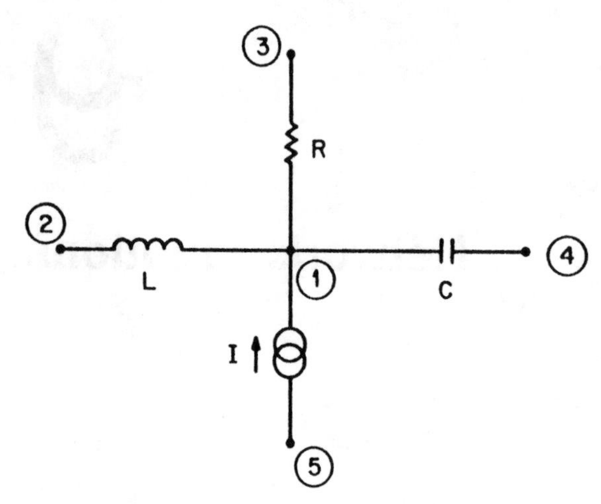

Figure 9.1 Circuit fragment for illustrating the KCL relationships.

unknown, and solving for them is the major purpose behind writing the system of nodal admittance equations for a circuit.) The current leaving node 1 via the capacitor is $(V_1 - V_4)sC$, the current leaving via the inductor is $(V_1 - V_2)/(sL)$, and the current leaving via the resistor is $(V_1 - V_3)/R$. The equation for node 1 can therefore be written as

$$(V_1 - V_4)sC + \frac{1}{R}(V_1 - V_3) + \frac{1}{sL}(V_1 - V_2) = I$$

or

$$\left(sC + \frac{1}{R} + \frac{1}{sL}\right)V_1 - sCV_2 - \frac{1}{R}V_3 - \frac{1}{sL}V_4 = I \tag{9.1}$$

Note that the terms *enter* and *leave* are somewhat relative. As depicted in Fig. 9.1, a current of I amperes (A) is entering node 1 from the independent current source. If the direction of the arrow were reversed, a current of I A would be leaving node 1 via the current source. However, we could also say that a current of $-I$ A would be entering the node. Likewise, currents actually entering a node via passive components can be thought of as negative currents leaving the node. In our discussions of nodal admittance equations, we will speak of currents *entering* nodes from current sources and currents *leaving* nodes via passive components—regardless of the actual direction of the current flows. The value will simply be negative if a current assumed to be entering is in fact leaving or if a current assumed to be leaving is actually entering.

An equation similar to (9.1) can be written for each node in a circuit, and the circuit is completely characterized by the set of such equations corresponding to all the ungrounded nodes.

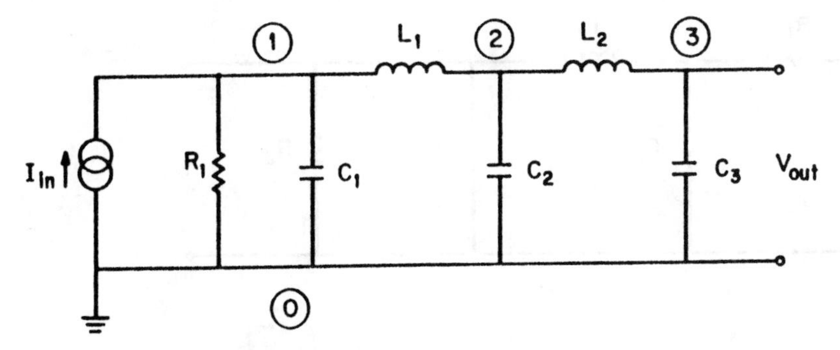

Figure 9.2 Circuit for Example 9.1.

Example 9.1 Find the set of nodal admittance equations for the circuit shown in Fig. 9.2. The node labeled 0 is grounded so the KCL equation for this node is not needed to characterize the circuit.

For node 1:

The current entering is I.
The current leaving via R_1 is V_1/R_1.
The current leaving via L_1 is $(V_1 - V_2)/(sL_1)$.
The current leaving via C_1 is $sC_1 V_1$.
Thus the KCL equation is given by

$$\frac{1}{R_1} V_1 + sC_1 V_1 + \frac{1}{sL_1}(V_1 - V_2) = I_{\text{in}}$$

or

$$\left(\frac{1}{R_1} + sC_1 + \frac{1}{sL_1}\right)V_1 - \frac{1}{sL_1} V_2 = I_{\text{in}} \tag{9.2}$$

For node 2:

The current entering is zero.
The current leaving via L_1 is $(V_2 - V_1)/(sL_1)$.
The current leaving via L_2 is $(V_2 - V_3)/(sL_2)$.
The current leaving via C_2 is $sC_2 V_2$.
Thus the KCL equation is given by

$$\frac{1}{sL_1}(V_2 - V_1) + \frac{1}{sL_2}(V_2 - V_3) + sC_2 V_2 = 0$$

or

$$-\frac{1}{sL_1} V_1 + \left(\frac{1}{sL_1} + \frac{1}{sL_2} + sC_2\right)V_2 - \frac{1}{sL_2} V_3 = 0 \tag{9.3}$$

For node 3:

The current entering is zero.
The current leaving via L_2 is $(V_3 - V_2)/(sL_2)$.
The current leaving via C_3 is $sC_3 V_3$.
Thus the KCL equation is given by

$$\frac{1}{sL_2}(V_3 - V_2) + sC_3 V_3 = 0$$

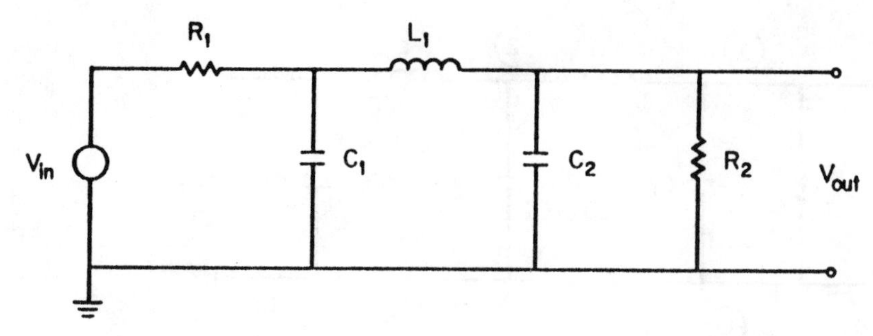

Figure 9.3 Circuit for Example 9.2.

or

$$-\frac{1}{sL_2}V_2+\left(\frac{1}{sL_2}+sC_3\right)V_3=0 \tag{9.4}$$

If a circuit to be analyzed contains any voltage sources, they must be converted to the corresponding Norton equivalent circuits before the nodal admittance equations can be formulated.

Example 9.2 Find the set of nodal admittance equations for the circuit shown in Fig. 9.3.

solution The voltage source can be eliminated by replacing the combination of the voltage source and resistor R_1 with the corresponding Norton equivalent (see Sec. 3.8) to yield the circuit shown in Fig. 9.4.

The equations for nodes 1 and 2, respectively, are given by

$$\left(\frac{1}{R_1}+sC_1+\frac{1}{sL_1}\right)V_1-\frac{1}{sL_1}V_2=\frac{V_{\text{in}}}{R_1} \tag{9.5}$$

$$-\frac{1}{sL_1}V_1+\left(\frac{1}{R_2}+\frac{1}{sL_1}+sC_2\right)V_2=0 \tag{9.6}$$

Subsequent steps in the computer-aided analysis of a circuit will be more convenient if the sets of equations such as given by (9.2) through (9.4) or (9.5)

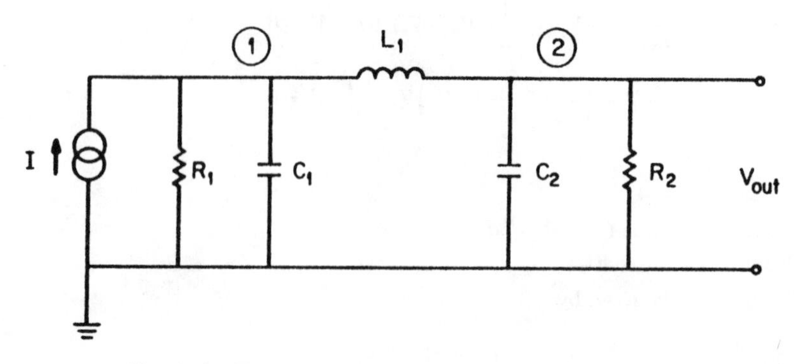

Figure 9.4 Circuit for Example 9.2 after modification to remove voltage source.

and (9.6) are put into the following matrix form:

$$\begin{bmatrix} y_{11} & y_{12} & \cdots & y_{1n} \\ y_{21} & y_{22} & \cdots & y_{2n} \\ \cdots & \cdots & \cdots & \cdots \\ y_{n1} & y_{n2} & \cdots & y_{nn} \end{bmatrix} \begin{bmatrix} V_1 \\ V_2 \\ \vdots \\ V_n \end{bmatrix} = \begin{bmatrix} J_1 \\ J_2 \\ \vdots \\ J_n \end{bmatrix} \tag{9.7}$$

where V_i = voltage at node i
J_i = current entering node i from current sources
y_{ij} = admittance which multiplies V_j in KCL equation for node i

All the entries in the second matrix of (9.7) are *node* voltages, and all the entries in the first matrix are branch *admittances*, hence Eq. (9.7) is often called the *nodal admittance form* of the network equations. If it is not important to depict individual matrix elements, Eq. (9.7) can be written as

$$\mathbf{Yv} = \mathbf{J}$$

It is customary to use uppercase bold roman letters for matrices (such as **Y**) and lowercase bold roman letters for column vectors (such as **v**). It seems wise to break with this custom as regards **J** since a lowercase j, whether boldface or lightface, would be too easily confused with either an index or the square root of -1.

Example 9.3 Put the nodal equations from Examples 9.1 and 9.2 into matrix form. Equations (9.2), (9.3), and (9.4) are expressed in matrix form as

$$\begin{bmatrix} \frac{1}{R_1} + sC_1 + \frac{1}{sL_1} & \frac{-1}{sL_1} & 0 \\ -\frac{1}{sL_1} & \frac{1}{sL_1} + \frac{1}{sL_2} + sC_2 & \frac{-1}{sL_2} \\ 0 & \frac{-1}{sL_2} & \frac{1}{sL_2} + sC_3 \end{bmatrix} \begin{bmatrix} V_1 \\ V_2 \\ V_3 \end{bmatrix} = \begin{bmatrix} I_{\text{in}} \\ 0 \\ 0 \end{bmatrix} \tag{9.8}$$

Equations (9.5) and (9.6) are expressed in matrix form as

$$\begin{bmatrix} \frac{1}{R_1} + sC_1 + \frac{1}{sL_1} & \frac{-1}{sL_1} \\ -\frac{1}{sL_1} & \frac{1}{sL_1} + sC_2 + \frac{1}{R_2} \end{bmatrix} \begin{bmatrix} V_1 \\ V_2 \end{bmatrix} = \begin{bmatrix} \frac{V_{\text{in}}}{R_1} \\ 0 \end{bmatrix} \tag{9.9}$$

So far, numerical values have purposely not been assigned to the various circuit elements so that the contributions of each element can be easily tracked through the various equation formulations. In a practical analysis problem, all the elements would be assigned values and we would solve the resulting system of equations to find particular node voltages or branch currents.

Example 9.4 The circuit shown in Fig. 9.4 can be used to implement a three-pole Chebyshev filter having a resistive load of R_2 and an input source comprising the current source and R_1. For typical resistor values of $R_1 = R_2 = 1\ \text{k}\Omega$ and a 3-dB frequency of 3 kHz, we need to assign $C_1 = C_2 = 0.1176\ \mu\text{F}$ and $L_1 = 57.74$ mH. The nodal admittance matrix for any particular steady-state frequency f can be obtained by substituting the given values into Eq. (9.9) with $s = j\omega = j2\pi f$. If we arbitrarily let $V_{\text{in}} = 1$ V, the independent current source must have $I_{\text{in}} = 1.0$ mA. By using the complex structure notation defined in App. C, the matrix equation for the circuit at a frequency of 3 kHz is given by

$$\begin{bmatrix} (1\text{e}-3, 1.298\text{e}-3) & (0.0, 9.188\text{e}-4) \\ (0.0, 9.188\text{e}-4) & (1\text{e}-3, 1.298\text{e}-3) \end{bmatrix} \begin{bmatrix} V_1 \\ V_2 \end{bmatrix} = \begin{bmatrix} (1\text{e}-3, 0.0) \\ (0.0, 0.0) \end{bmatrix} \tag{9.10}$$

Just for the sake of comparison, the corresponding equation for $f = 30$ Hz is given by

$$\begin{bmatrix} (1\text{e}-3, -9.186\text{e}-2) & (0.0, 9.188\text{e}-2) \\ (0.0, 9.188\text{e}-2) & (1\text{e}-3, -9.186\text{e}-2) \end{bmatrix} \begin{bmatrix} V_1 \\ V_2 \end{bmatrix} = \begin{bmatrix} (1\text{e}-3, 0.0) \\ (0.0, 0.0) \end{bmatrix} \tag{9.11}$$

Since a voltage-controlled current source (VCCS) is defined by its transfer admittance, it is a straightforward matter to include VCCSs in the nodal admittance formulation. The circuit configuration for a VCCS or voltage-to-current transducer (VCT) is shown in Fig. 9.5. The presence of a VCCS in a circuit causes a current of $I = gV$ to leave node j and enter node k. The size of this current can be expressed as a function of the node voltage at nodes h and i:

$$\text{Current leaving node } j = gV_{nh} - gV_{ni}$$

$$\text{Current leaving node } k = gV_{ni} - gV_{nh}$$

9.2 Construction Rules for the Nodal Admittance Formulation

The algebraic manipulations used to obtain the matrix form given in (9.7) can be avoided by direct application of the following rules.

Rule 9.1 (Conductances in the nodal admittance matrix). The effects of a conductance G connected between nodes j and k can be incorporated into the nodal admittance matrix via the following steps:

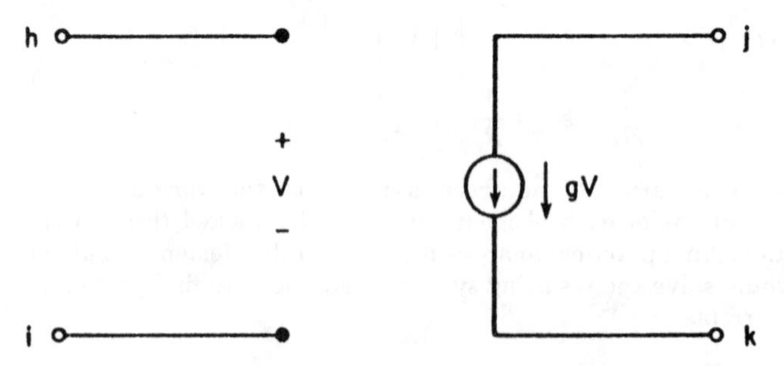

Figure 9.5 Circuit configuration for a voltage-controlled current source.

1. Add the value G to the element in row j, column j of $\mathbf{Y}_n$.
2. Add the value G to the element in row k, column k of $\mathbf{Y}_n$.
3. Subtract the value G from the element in row j, column k of $\mathbf{Y}_n$.
4. Subtract the value G from the element in row k, column j of $\mathbf{Y}_n$.

Rule 9.2 (Resistors in the nodal admittance matrix). The effects of a resistance R connected between nodes j and k can be incorporated into the nodal admittance matrix via the following steps:

1. Add the value $1/R$ to the element in row j, column j of $\mathbf{Y}_n$.
2. Add the value $1/R$ to the element in row k, column k of $\mathbf{Y}_n$.
3. Subtract the value $1/R$ from the element in row j, column k of $\mathbf{Y}_n$.
4. Subtract the value $1/R$ from the element in row k, column j of $\mathbf{Y}_n$.

Rule 9.3 (Capacitors in the nodal admittance matrix). At a frequency f, the sinusoidal steady-state effects of a capacitance C connected between nodes j and k can be incorporated into the nodal admittance matrix via the following steps:

1. Add the value $sC = 2\pi fC$ to the element in row j, column j of $\mathbf{Y}_n$.
2. Add the value $sC = 2\pi fC$ to the element in row k, column k of $\mathbf{Y}_n$.
3. Subtract the value $sC = 2\pi fC$ from the element in row j, column k of $\mathbf{Y}_n$.
4. Subtract the value $sC = 2\pi fC$ from the element in row k, column j of $\mathbf{Y}_n$.

Rule 9.4 (Inductors in the nodal admittance matrix). At a frequency f, the sinusoidal steady-state effects of an inductance L connected between nodes j and k can be incorporated into the nodal admittance matrix via the following steps:

1. Add the value $1/(2\pi fL)$ to the element in row j, column j of $\mathbf{Y}_n$.
2. Add the value $1/(2\pi fL)$ to the element in row k, column k of $\mathbf{Y}_n$.
3. Subtract the value $1/(2\pi fL)$ from the element in row j, column k of $\mathbf{Y}_n$.
4. Subtract the value $1/(2\pi fL)$ from the element in row k, column j of $\mathbf{Y}_n$.

Rule 9.5 (Current sources in the nodal admittance matrix formulation). The effects of an independent current source J connected from node j to node k can be incorporated into the source vector $\mathbf{J}$ of the nodal admittance matrix equation via the following steps:

1. Add the value J to element k of $\mathbf{J}$.
2. Subtract the value J from element j of $\mathbf{J}$.

Rule 9.6 (Voltage-controlled current sources in the nodal admittance matrix). The effects of a voltage-controlled current source having a transfer admittance of g and connected as shown in Fig. 9.5 can be incorporated into the nodal admittance matrix via the following steps:

1. Add the value g to the element in row j, column h of $\mathbf{Y}_n$.
2. Add the value g to the element in row k, column i of $\mathbf{Y}_n$.
3. Subtract the value g from the element in row j, column i of $\mathbf{Y}_n$.
4. Subtract the value g from the element in row k, column h of $\mathbf{Y}_n$.

A C function **BuildNodalEqn( )**, which is based upon Rules 9.1 through 9.6, is shown in Listing 9.1. In the algebraic approach of Sec. 9.1, we excluded the grounded "zero node" from the $\mathbf{Y}_n$ matrix. However, the program provided generates a $\mathbf{Y}_n$ matrix that includes all the specified nodes, including node 0 which is assumed to be grounded. Since C uses "zero origin indexing," it is convenient to form the complete matrix and ignore row 0 and column 0 in subsequent processing steps.

```
/***********************************************/
/*                                             */
/*   Listing 9.1                               */
/*                                             */
/*   BuildNodalEqn()                           */
/*                                             */
/***********************************************/
#include "globDefs.h"
#include "protos.h"
int BuildNodalEqn(    real freq,
                      struct complex Y[][MAX_COLUMNS],
                      struct complex J_vector[],
                      int *numNodes)

{
enum componentTypes component;
real value1,value2, value3;
real yNew;
real omega;
char depVar[5], depVar2[5];
int hNode, iNode, jNode, kNode;
int maxNode;
int i,j;
logical done;
/*--------------------------------------*/
/*  zero out the Y matrix and J vector   */
```

```
for(i=0; i<MAX_COLUMNS; i++)  J_vector[i] = cmplx(0.0,0.0);
for(i=0; i<MAX_COLUMNS; i++)
    {
    for(j=0; j<MAX_COLUMNS; j++) Y[i][j] = cmplx(0.0,0.0);
    }
maxNode = 0;
omega = 2.0 * PI * freq;
done = FALSE;
for(;;)
    {
    GetBranchData(&component, &value1, &value2, &value3,&hNode,
                   &iNode, &jNode, &kNode, &done, depVar, depVar2);
    if(jNode > maxNode) maxNode = jNode;
    if(kNode > maxNode) maxNode = kNode;
    if(hNode > maxNode) maxNode = hNode;
    if(iNode > maxNode) maxNode = iNode;
    switch (component)
        {
        case conductance:        /*  see Rule 9.1  */
            Y[jNode][jNode].Re += value1;
            Y[kNode][kNode].Re += value1;
            Y[jNode][kNode].Re -= value1;
            Y[kNode][jNode].Re -= value1;

        case resistor:               /*  see Rule 9.2  */
            yNew = 1.0/value1;
            Y[jNode][jNode].Re += yNew;
            Y[kNode][kNode].Re += yNew;
            Y[jNode][kNode].Re -= yNew;
            Y[kNode][jNode].Re -= yNew;
            break;
case capacitor:                  /*  see Rule 9.3  */
    yNew = omega * value1;
    Y[jNode][jNode].Im += yNew;
    Y[kNode][kNode].Im += yNew;
    Y[jNode][kNode].Im -= yNew;
    Y[kNode][jNode].Im -= yNew;
    break;

case inductor:                   /*  see Rule 9.4  */
    yNew = -1.0/(omega * value1);
    Y[jNode][jNode].Im += yNew;
    Y[kNode][kNode].Im += yNew;
    Y[jNode][kNode].Im -= yNew;
```

```
Y[kNode][jNode].Im -= yNew;
break;
case currentSource:              /* see Rule 9.5 */
    J_vector[jNode].Re -= value1;
    J_vector[kNode].Re += value1;
    break;
        case VCCS:                      /* see Rule 9.6 */
            Y[jNode][hNode].Re += value1;
            Y[kNode][iNode].Re += value1;
            Y[jNode][iNode].Re -= value1;
            Y[kNode][hNode].Re -= value1;

        default:
            return -1;
        }
    if(done) break;
    }
*numNodes = maxNode;
return 0;
}
```

9.3 Tableau Formulation

The tableau formulation of a network's equations includes the KCL, KVL, and constitutive equations. The KCL equations for a network of b branches and $n+1$ nodes can be expressed in matrix form as

$$\mathbf{A}\mathbf{I}_b = \mathbf{0} \tag{9.12}$$

where $\mathbf{A}$ = n-row, b-column reduced incidence matrix defined in Sec. 8.2 (The matrix is identified as "reduced" because the row corresponding to the reference node is omitted.)
$\mathbf{I}_b$ = b-element column vector of branch currents

The KVL equations can be expressed in matrix form as

$$\mathbf{A}^t\mathbf{V}_n = \mathbf{V}_b \qquad \text{or} \qquad \mathbf{V}_b - \mathbf{A}^t\mathbf{V}_n = \mathbf{0} \tag{9.13}$$

where $\mathbf{A}^t$ = b-row, n-column transpose of $\mathbf{A}$
$\mathbf{V}_n$ = n-element column vector of node voltages
$\mathbf{V}_b$ = b-element column vector of branch voltages

The constitutive equations for the elements within the network can be expressed in matrix form as

$$\mathbf{Y}_b\mathbf{V}_b + \mathbf{Z}_b\mathbf{I}_b = \mathbf{W}_b \tag{9.14}$$

where $\mathbf{Y}_b = b$-row, b-column branch admittance matrix
$\mathbf{Z}_b = b$-row, b-column branch impedance matrix
$\mathbf{W}_b = b$-element column vector that includes any independent current sources and/or independent voltage sources

The tableau formulation incorporates Eqs. (9.12) through (9.14) into the matrix equation

$$\begin{bmatrix} \mathbf{1} & \mathbf{0} & -\mathbf{A}^t \\ \mathbf{Y}_b & \mathbf{Z}_b & \mathbf{0} \\ \mathbf{0} & \mathbf{A} & \mathbf{0} \end{bmatrix} \begin{bmatrix} \mathbf{V}_b \\ \mathbf{I}_b \\ \mathbf{V}_n \end{bmatrix} = \begin{bmatrix} \mathbf{0} \\ \mathbf{W}_b \\ \mathbf{0} \end{bmatrix} \tag{9.15}$$

where $\mathbf{1}$ and $\mathbf{0}$ denote the (appropriately sized) identity matrix and zero matrix, respectively. At times it will be convenient to refer to the first matrix on the left-hand side (LHS) of (9.15) as the $\mathbf{T}$ matrix. For a given network, evaluation of submatrices $\mathbf{A}$, $\mathbf{Y}_b$, $\mathbf{Z}_b$, and $\mathbf{W}_b$ is straightforward. Equation (9.15) can then be solved to obtain $\mathbf{V}_b$, $\mathbf{I}_b$, and $\mathbf{V}_n$. The specific steps involved in the generation of the required matrices are detailed in Algorithm 9.1.

Algorithm 9.1 (Generation of the tableau matrix equations)

Step 1. Number each node in the circuit, beginning with 0 for the reference (i.e., grounded) node.

Step 2. Use the rules given in Sec. 8.1 to construct the oriented graph for the circuit. Number each edge in the graph, using some scheme that will avoid confusion between node numbers and edge (branch) numbers. (In this book, node numbers are always circled in illustrations and preceded by n in text. Thus, v_2 denotes the branch voltage for edge 2 while V_{n2} denotes the voltage on node 2 relative to the reference node.)

Step 3. Working from the oriented graph, construct the reduced incidence matrix $\mathbf{A}$. For each edge j that begins at (i.e., points away from) node i, set $a_{ij} = 1$. For each edge j that ends at (i.e., points toward) node i, set $a_{ij} = -1$.

Step 4. Form the $\mathbf{Y}_b$, $\mathbf{Z}_b$, and $\mathbf{W}_b$ matrices, using entries as indicated in Tables 9.1 and 9.2.

TABLE 9.1 Tableau Entries for One-Port Elements in Branch *i*

Resistor	Inductance	Capacitance
$y_{ii} = 1$ $z_{ii} = -R$	$y_{ii} = 1$ $z_{ii} = -sL$	$y_{ii} = sC$ $z_{ii} = -1$
Conductance	**Independent voltage source**	**Independent current source**
$y_{ii} = G$ $z_{ii} = -1$	$y_{ii} = 1$ $w_i = E$	$z_{ii} = 1$ $w_i = J$

TABLE 9.2 Tableau Entries for Two-Ports with Input in Branch *i* and Output in Branch *j*

VCCS	VCVS	CCCS	CCVS	Ideal operational amplifier
$y_{ji} = g$ $z_{ii} = 1$ $z_{jj} = -1$	$y_{ji} = \mu$ $y_{jj} = -1$ $z_{ii} = 1$	$y_{ii} = 1$ $z_{jj} = -1$ $z_{ji} = \alpha$	$y_{ii} = 1$ $y_{jj} = -1$ $z_{ji} = \gamma$	$y_{ii} = 1$ $z_{ji} = 1$

Step 5. In the positions indicated by Eq. (9.15), catenate the various submatrices that were formed in steps 3 and 4. ■

Example 9.5 Generate the tableau for the circuit shown in Fig. 9.4.

solution The nodes have already been numbered in the figure. The corresponding oriented graph is shown in Fig. 9.6, and matrices $\mathbf{A}$ and $-\mathbf{A}^t$ are given by

$$\mathbf{A} = \begin{bmatrix} -1 & 1 & 1 & 1 & 0 & 0 \\ 0 & 0 & 0 & -1 & 1 & 1 \end{bmatrix} \qquad -\mathbf{A}^t = \begin{bmatrix} 1 & 0 \\ -1 & 0 \\ -1 & 0 \\ -1 & 1 \\ 0 & -1 \\ 0 & -1 \end{bmatrix}$$

Building $\mathbf{Y}_b$, $\mathbf{Z}_b$, and $\mathbf{W}_b$ with the appropriate entries from Table 9.1, we obtain the tableau formulation

$$\left[\begin{array}{cccccc|cccccc|cc}
1 & & & & & & & & & & & & 1 & 0 \\
 & 1 & & & & & & & & & & & -1 & 0 \\
 & & 1 & & & & & & & & & & -1 & 0 \\
 & & & 1 & & & & & \mathbf{0} & & & & -1 & 1 \\
 & & & & 1 & & & & & & & & 0 & -1 \\
 & & & & & 1 & & & & & & & 0 & -1 \\
\hline
0 & & & & & & 1 & & & & & & & \\
 & 1 & & & & & & -R_1 & & & & & & \\
 & & sC_1 & & & & & & -1 & & & & & \\
 & & & 1 & & & & & & -sL & & & \mathbf{0} & \\
 & & & & sC_2 & & & & & & -1 & & & \\
 & & & & & 1 & & & & & & -R_2 & & \\
\hline
 & & \mathbf{0} & & & & -1 & 1 & 1 & 1 & 0 & 0 & \mathbf{0} & \\
 & & & & & & 0 & 0 & 0 & -1 & 1 & 1 & &
\end{array}\right]
\begin{bmatrix} V_1 \\ V_2 \\ V_3 \\ V_4 \\ V_5 \\ V_6 \\ I_1 \\ I_2 \\ I_3 \\ I_4 \\ I_5 \\ I_6 \\ V_{n1} \\ V_{n2} \end{bmatrix}
=
\left[\begin{array}{c} \\ \\ \\ \mathbf{0} \\ \\ \\ \hline V_{\text{in}}/R_1 \\ 0 \\ 0 \\ 0 \\ 0 \\ 0 \\ \hline \mathbf{0} \\ \\ \end{array}\right]$$

The $\mathbf{T}$ matrix is the biggest, baddest, most complete formulation possible; but since the matrix will contain $(2b + n)^2$ entries, it is clear that sparse matrix solution techniques will be necessary for all but the smallest networks. Another, more manageable approach is presented in Sec. 9.4.

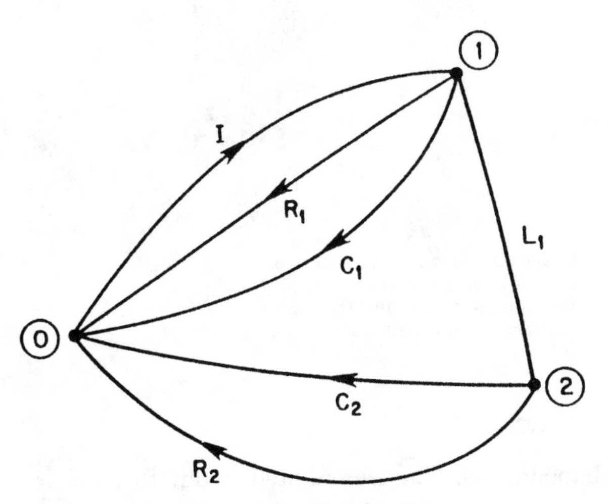

Figure 9.6 Oriented graph for Example 9.5.

9.4 Augmented Nodal Formulation

The nodal admittance formulation, as presented so far, is limited to circuit elements that can be modeled in terms of admittances and current sources. The nodal admittance matrix and node-voltage vector can be augmented to allow inclusion of other types of circuit elements. The rows and columns added to the nodal admittance matrix will contain elements which are not admittances, so the augmented matrix will be denoted by **T** with **Y** being reserved for the nodal admittance submatrix comprising the upper left-hand partition in **T**. Likewise, the rows added to **v** will contain elements that are not node voltages, and the rows added to **J** will contain elements that are not independent currents. Therefore, the vector of dependent variables will be denoted by **x** with **v** reserved for the node-voltage submatrix comprising the upper partition in **x**. The vector of independent sources will be denoted by **W** and **J** reserved for the submatrix of independent currents comprising the upper partition in **W**. The relationships between **T**, **Y**, **x**, **v**, **W**, and **J** are summarized by the following matrix equation:

$$\underbrace{\left[\begin{array}{c:c} \mathbf{Y} & \begin{array}{c}\text{added}\\ \text{columns}\end{array} \\ \hdashline \multicolumn{2}{c}{\begin{array}{c}\text{added}\\ \text{rows}\end{array}} \end{array}\right]}_{\mathbf{T}} \underbrace{\left[\begin{array}{c} \mathbf{v} \\ \hdashline \text{added}\\ \text{rows} \end{array}\right]}_{\mathbf{x}} = \underbrace{\left[\begin{array}{c} \mathbf{J} \\ \hdashline \text{added}\\ \text{rows} \end{array}\right]}_{\mathbf{W}}$$

Impedances. In the nodal admittance formulation, we took the impedances of inductors and resistors and converted them to corresponding admittances. For some components, such as current-controlled resistors, it will be neces-

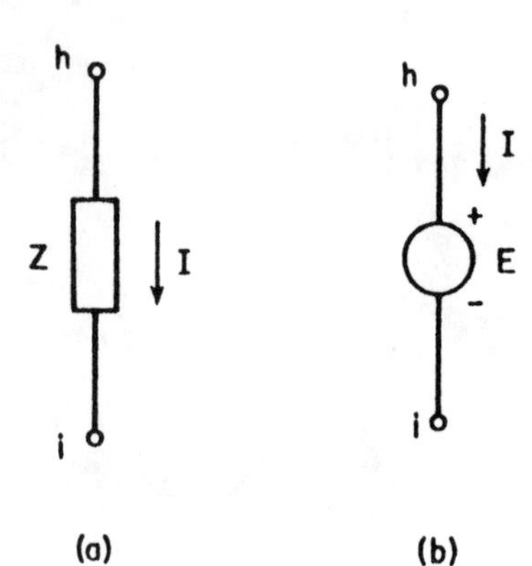

Figure 9.7 Circuit configuration for (*a*) impedances and (*b*) voltage sources in the augmented nodal admittance formulation.

sary to work directly with impedances. The augmented nodal formulation provides a means to do this. Consider the impedance depicted in Fig. 9.7*a*. A current I leaves node h and enters node i. The voltage drop caused by the current through the impedance is simply the difference between the two node voltages:

$$I_h = I \qquad I_i = -I$$

$$V_h - V_i = zI_z$$

Rule 9.7 (Impedances in the augmented nodal formulation). The effects of an impedance connected as shown in Fig. 9.7*a* can be incorporated into the system matrix equation via the following steps:

1. Append a new column to the right-hand side of system matrix **T**.
2. Set the element in row j of the new column equal to 1.
3. Set the element in row k of the new column equal to -1.
4. Append a new row to the bottom of system matrix **T**.
5. Put the value of $-z$ into the element where the new row intersects the new column.
6. Set the element in column j of the new row equal to 1.
7. Set the element in column k of the new row equal to -1.
8. Append a new element to the bottom of column vector **x**. Associate this element with the current denoted by I in Fig. 9.7*a*.

The results of steps 1 through 8 are summarized in entry (1) of Table 9.3.

If it is ever necessary to model two separate nodes connected by a short circuit, we can use Rule 9.7 with $z = 0$.

TABLE 9.3 Matrix Entries for the Augmented Nodal Formulation

Impedance:

$$
\begin{array}{c} \\ h \\ i \\ m+1 \end{array}
\begin{array}{c}
\begin{array}{ccc} h & i & m+1 \end{array} \\
\left[\begin{array}{cc|c} & & 1 \\ & & -1 \\ \hline 1 & -1 & -z \end{array}\right]
\end{array}
\left[\begin{array}{c} V_{nh} \\ V_{ni} \\ \hline I \end{array}\right]
=
\left[\begin{array}{c} \\ \\ \\ \end{array}\right]
\qquad (1)
$$

Independent voltage source:

$$
\begin{array}{c} \\ h \\ i \\ m+1 \end{array}
\begin{array}{c}
\begin{array}{ccc} h & i & m+1 \end{array} \\
\left[\begin{array}{cc|c} & & 1 \\ & & -1 \\ \hline 1 & -1 & \end{array}\right]
\end{array}
\left[\begin{array}{c} V_{nh} \\ V_{ni} \\ \hline I \end{array}\right]
=
\left[\begin{array}{c} \\ \\ \hline E \end{array}\right]
\qquad (2)
$$

Voltage-controlled voltage source:

$$
\begin{array}{c} \\ h \\ i \\ j \\ k \\ m+1 \end{array}
\begin{array}{c}
\begin{array}{ccccc} h & i & j & k & m+1 \end{array} \\
\left[\begin{array}{cccc|c} & & & & \\ & & & & \\ & & & & 1 \\ & & & & -1 \\ \hline -\mu & \mu & 1 & -1 & \end{array}\right]
\end{array}
\left[\begin{array}{c} V_{nh} \\ V_{ni} \\ V_{nj} \\ V_{nk} \\ \hline I \end{array}\right]
=
\left[\begin{array}{c} \\ \\ \\ \\ \\ \end{array}\right]
\qquad (3)
$$

Current-controlled current source:

$$
\begin{array}{c} \\ h \\ i \\ j \\ k \\ m+1 \end{array}
\begin{array}{c}
\begin{array}{ccccc} h & i & j & k & m+1 \end{array} \\
\left[\begin{array}{cccc|c} & & & & 1 \\ & & & & -1 \\ & & & & \alpha \\ & & & & -\alpha \\ \hline 1 & -1 & & & \end{array}\right]
\end{array}
\left[\begin{array}{c} V_{nh} \\ V_{ni} \\ V_{nj} \\ V_{nk} \\ \hline I \end{array}\right]
=
\left[\begin{array}{c} \\ \\ \\ \\ \\ \end{array}\right]
\qquad (4)
$$

Current-controlled voltage source:

$$
\begin{array}{c} \\ h \\ i \\ j \\ k \\ m+1 \\ m+2 \end{array}
\begin{array}{c}
\begin{array}{cccccc} h & i & j & k & m+1 & m+2 \end{array} \\
\left[\begin{array}{cccc|c|c} & & & & 1 & \\ & & & & -1 & \\ & & & & & 1 \\ & & & & & -1 \\ \hline 1 & -1 & & & & \\ \hline & & 1 & -1 & -\gamma & \end{array}\right]
\end{array}
\left[\begin{array}{c} V_{nh} \\ V_{ni} \\ V_{nj} \\ V_{nk} \\ \hline I_1 \\ \hline I_2 \end{array}\right]
=
\left[\begin{array}{c} \\ \\ \\ \\ \\ \\ \end{array}\right]
\qquad (5)
$$

TABLE 9.3 (continued)

Transformer:

$$
\begin{array}{c}
\\ h \\ i \\ j \\ k \\ m+1 \\ m+2
\end{array}
\begin{array}{c}
\begin{array}{cccccc} h & i & j & k & m+1 & m+2 \end{array} \\
\left[\begin{array}{cccc|c|c}
 & & & & 1 & \\
 & & & & -1 & \\
 & & & & & 1 \\
 & & & & & -1 \\
\hline
1 & -1 & & & -sL_1 & -sM \\
\hline
 & & 1 & -1 & -sM & -sL_2
\end{array}\right]
\end{array}
\left[\begin{array}{c}
V_{nh} \\ V_{ni} \\ V_{nj} \\ V_{nk} \\ \hline I_1 \\ \hline I_2
\end{array}\right]
=
\left[\begin{array}{c}
\\ \\ \\ \\ \\ \\
\end{array}\right]
\qquad (6)
$$

Operational amplifier:

$$
\begin{array}{c}
\\ h \\ i \\ j \\ k \\ m+1
\end{array}
\begin{array}{c}
\begin{array}{ccccc} h & i & j & k & m+1 \end{array} \\
\left[\begin{array}{cccc|c}
 & & & & \\
 & & & & \\
 & & & & 1 \\
 & & & & -1 \\
\hline
1 & -1 & & &
\end{array}\right]
\end{array}
\left[\begin{array}{c}
V_{nh} \\ V_{ni} \\ V_{nj} \\ V_{nk} \\ \hline I
\end{array}\right]
=
\left[\begin{array}{c}
\\ \\ \\ \\ \\
\end{array}\right]
\qquad (7)
$$

Independent voltage sources. In the nodal admittance formulation, it was necessary to convert an independent voltage source plus series resistance to the equivalent independent current source plus shunt conductance. The augmented nodal formulation provides a means for direct handling of voltage sources without the need to convert them to current sources. Consider the voltage source depicted in Fig. 9.7*b*. A current I leaves node h and enters node i. The source voltage is simply the difference between the two node voltages:

$$I_h = I \qquad I_i = -I$$

$$V_h - V_i = E$$

Rule 9.8 (Voltage sources in the augmented nodal formulation). The effects of an independent voltage source connected as shown in Fig. 9.7*b* can be incorporated into the system matrix equation via the following steps:

1. Append a new column to the right-hand side of system matrix **T**.
2. Set the element in row j of the new column equal to 1.
3. Set the element in row k of the new column equal to -1.
4. Append a new row to the bottom of system matrix **T**.
5. Set the element in column j of the new row equal to 1.
6. Set the element in column k of the new row equal to -1.

7. Append a new element to the bottom of source vector **W**. Set this element equal to source voltage E.
8. Append a new element to the bottom of column vector **x**. Associate this element with the current denoted by I in Fig. 9.7*b*.

The results of steps 1 through 8 are summarized in entry (2) of Table 9.3.

Voltage-controlled voltage sources. The circuit configuration for a voltage-controlled voltage source (VCVS) or voltage-to-voltage transducer (VVT) is shown in Fig. 9.8. As depicted in the figure, the VCVS presents an open circuit to the nodes labeled h and i, and therefore it has no impact on the current balances at these nodes. However, the presence of the VCVS does cause a current (labeled I) to leave node j and enter node k. This current is incorporated into the system matrix equation via steps 1 through 4 of Rule 9.9. The output voltage appearing between nodes j and k is defined to be μ times the input voltage appearing between nodes h and i. Thus,

$$\mu(V_{nh} - V_{ni}) = V_{nj} - V_{nk}$$

or

$$-\mu V_{nh} + \mu V_{ni} + V_{nj} - V_{nk} = 0 \tag{9.16}$$

Equation (9.16) is incorporated into the system equation via steps 5 through 9 of Rule 9.9.

Rule 9.9 (Voltage-controlled voltage sources in the augmented nodal formulation). The effects of a VCVS configured as shown in Fig. 9.8 can be incorporated into the system matrix equation via the following steps:

1. Append a new column to the right-hand side of system matrix **T**.
2. Set the element in row j of the new column equal to 1.
3. Set the element in row k of the new column equal to -1.
4. Append a new element to the bottom of column vector **x**. Associate this element with the current denoted by I in Fig. 9.8.

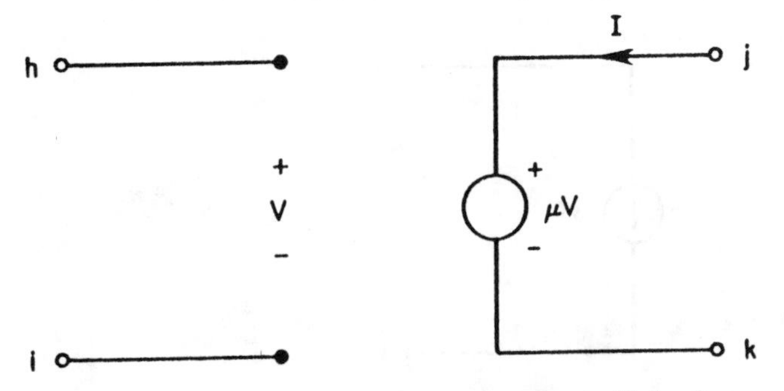

Figure 9.8 Circuit configuration for a voltage-controlled voltage source.

5. Append a new row to the bottom of system matrix **T**.
6. Set the element in column h of the new row equal to $-\mu$.
7. Set the element in column i of the new row equal to μ.
8. Set the element in column j of the new row equal to 1.
9. Set the element in column k of the new row equal to -1.

The results of steps 1 through 9 are summarized in entry (3) of Table 9.3.

Current-controlled current source. The circuit configuration for a current-controlled current source (CCCS) or current-to-current transducer (CCT) is shown in Fig. 9.9. The presence of the CCCS in a circuit causes a current I to leave node h and enter node i. In addition, a current of αI leaves node j and enters node k.

Rule 9.10 (Current-controlled current sources in the augmented nodal formulation). The effects of a CCCS configured as shown in Fig. 9.9 can be incorporated into the system matrix equation via the following steps:

1. Append a new column to the right-hand side of system matrix **T**.
2. Set the element in row h of the new column equal to 1.
3. Set the element in row i of the new column equal to -1.
4. Set the element in row j of the new column equal to α.
5. Set the element in row k of the new column equal to $-\alpha$.
6. Append a new row to the bottom of system matrix **T**.
7. Set the element in column h of the new row equal to 1.
8. Set the element in column i of the new row equal to -1.
9. Append a new element to the bottom of the column vector **x**. Associate this element with the current denoted by I in Fig. 9.9.

The results of steps 1 through 9 are summarized in entry (4) of Table 9.3.

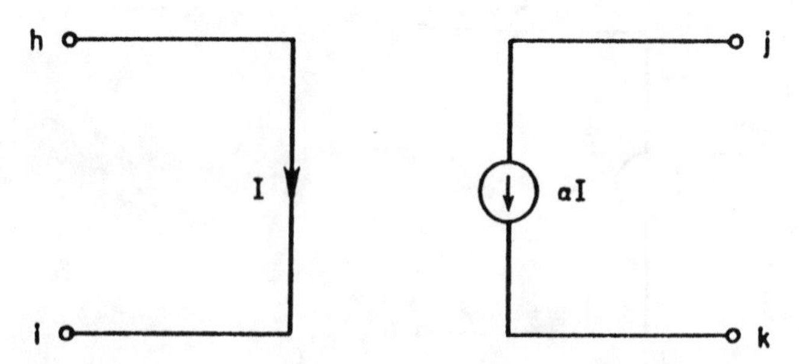

Figure 9.9 Circuit configuration for a current-controlled current source.

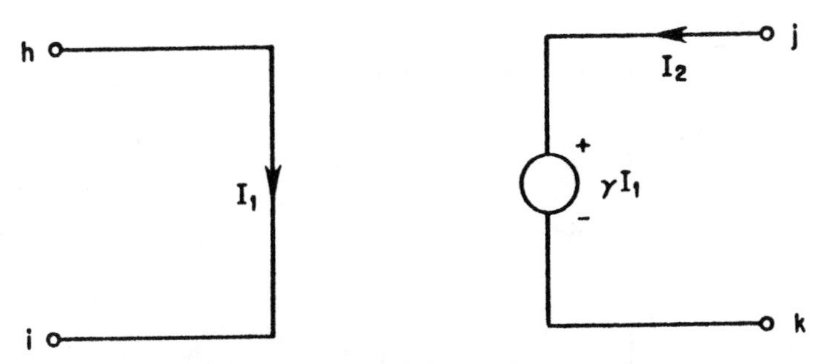

Figure 9.10 Circuit configuration for a current-controlled voltage source.

Current-controlled voltage source. The circuit configuration for a current-controlled voltage source (CCVS) or current-to-voltage transducer (CVT) is shown in Fig. 9.10. The presence of the CCVS in a circuit causes a current of I_1 to leave node h and enter node i. In addition, a current I_2 leaves node j and enters node k. The voltage drop from node j to node k is equal to γI_1.

$$V_{nj} - V_{nk} = \gamma I_1$$

or

$$V_{nj} - V_{nk} - \gamma I_1 = 0$$

Rule 9.11 (Current-controlled voltage sources in the augmented nodal formulation). The effects of a CCVS configured as shown in Fig. 9.10 can be incorporated into the system matrix equation via the following steps:

1. Append a new row to the bottom of system matrix **T**.
2. Set the element in column h of the new row equal to 1.
3. Set the element in column i of the new row equal to -1.
4. Append a second new row to the bottom of system matrix **T**.
5. Set the element in column j of the second new row equal to 1.
6. Set the element in column k of the second new row equal to -1.
7. Append a new column to the right-hand side of system matrix **T**.
8. Put the value of $-\gamma$ into the element where the new column intersects the new row that was appended in step 4 (i.e., the second new row).
9. Set the element in row h of the new column equal to 1.
10. Set the element in row i of the new column equal to -1.
11. Append a second new column to the right-hand side of system matrix **T**.
12. Set the element in row j of the second new column equal to 1.
13. Set the element in row k of the second new column equal to -1.
14. Append a new element to the bottom of column vector **x**. Associate this element with the current denoted by I in Fig. 9.10.

The results of steps 1 through 14 are summarized in entry (5) of Table 9.3.

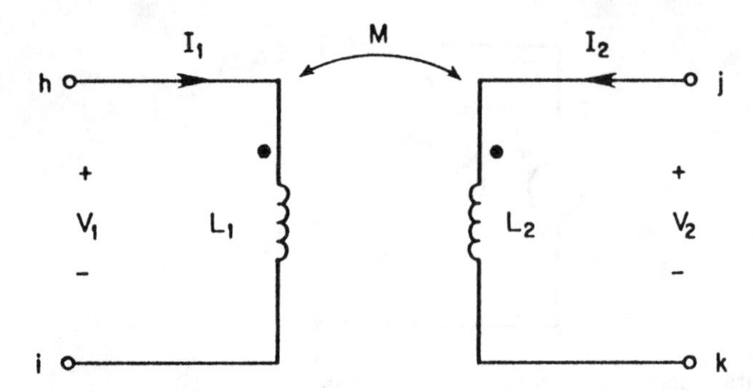

Figure 9.11 Circuit configuration for a transformer in the augmented nodal admittance formulation.

Transformers. The circuit configuration for a transformer is shown in Fig. 9.11. The mutual inductance between L_1 and L_2 is denoted as M. The presence of the transformer in a circuit causes a current of I_1 to leave node h and enter node i. In addition, a current I_2 leaves node j and enters node k. Based upon the results of Sec. 2.5, we can write

$$V_{nh} - V_{ni} = sL_1I_1 + sMI_2 \tag{9.17}$$

$$V_{nj} - V_{nk} = sMI_1 + sL_2I_2 \tag{9.18}$$

Equations (9.17) and (9.18) can be incorporated into the system matrix equation by using Rule 9.12.

Rule 9.12 (Transformers in the augmented nodal formulation). The effects of a transformer configured as shown in Fig. 9.11 can be incorporated into the system matrix equation via the following steps:

1. Append a new row to the bottom of system matrix **T**.
2. Set the element in column h of the new row equal to 1.
3. Set the element in column i of the new row equal to -1.
4. Append a second new row to the bottom of system matrix **T**.
5. Set the element in column j of the second new row equal to 1.
6. Set the element in column k of the second new row equal to -1.
7. Append a new column to the right-hand side of system matrix **T**.
8. Put the value of $-sL_1$ into the element where the new column intersects the new row that was appended in step 1 (i.e., the first new row).
9. Put the value of $-sM$ into the element where the new column intersects the new row that was appended in step 4 (i.e., the second new row).
10. Set the element in row h of the new column equal to 1.

11. Set the element in row i of the new column equal to -1.
12. Append a second new column to the right-hand side of system matrix $\mathbf{T}$.
13. Put the value of $-sM$ into the element where the second new column intersects the first new row.
14. Put the value of $-sL_2$ into the element where the second new column intersects the second new row.
15. Set the element in row j of the second new column equal to 1.
16. Set the element in row k of the second new column equal to -1.
17. Append a new element to the bottom of column vector $\mathbf{x}$. Associate this element with the current denoted by I_1 in Fig. 9.11.
18. Append a second new element to the bottom of column vector $\mathbf{x}$. Associate this element with the current denoted by I_2 in Fig. 9.11.

The results of steps 1 through 18 are summarized in entry (6) of Table 9.3.

Operational amplifiers. The circuit connections for an ideal operational amplifier (op amp) are shown in Fig. 9.12.

Rule 9.13 (Ideal op amps in the augmented nodal formulation). The effects of an ideal operational amplifier configured as shown in Fig. 9.12 can be incorporated into the system matrix equation via the following steps:

1. Append a new row to the bottom of system matrix $\mathbf{T}$.
2. Set the element in column h of the new row equal to 1.
3. Set the element in column i of the new row equal to -1.
4. Append a new column to the right-hand side of system matrix $\mathbf{T}$.
5. Set the element in row j of the new column equal to 1.
6. Set the element in row k of the new column equal to -1.
7. Append a new element to the bottom of column vector $\mathbf{x}$. Associate this element with the current denoted by I in Fig. 9.12.

The results of steps 1 through 7 are summarized in entry (7) of Table 9.3.

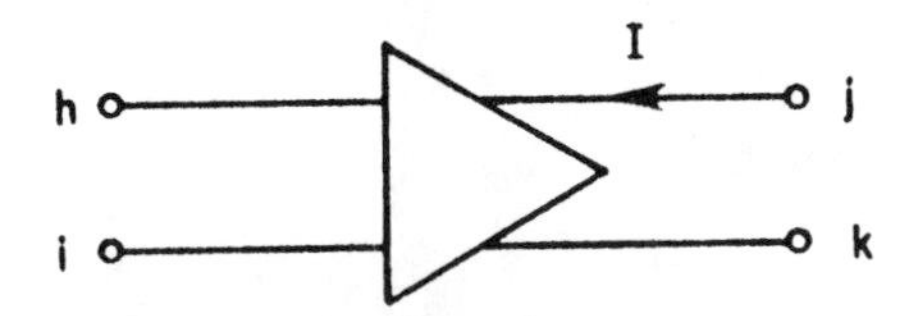

Figure 9.12 Circuit configuration for an operational amplifier in the augmented nodal admittance formulation.

Algorithm 9.2 (Generation of the augmented nodal admittance formulation)

Step 1. For the capacitors, conductances, and voltage-controlled current sources in the circuit, use Rules 9.1, 9.3, and 9.5 to generate the nodal admittance partition **Y** of system matrix **T**.

Step 2. For the independent current sources in the circuit, use Rule 9.6 to generate the **J** partition of the **W** vector.

Step 3. Incorporate impedances, independent voltage sources, voltage-controlled voltage sources, current-controlled voltage sources, current-controlled current sources, operational amplifiers, and transformers into the system matrix, using Rules 9.7 through 9.13 as appropriate. ■

Listing 9.2 contains the **C** function **BuildAugNodalEqn( )**, which generates the augmented nodal admittance matrix. Listing 9.3 contains the **C** function **SetFreq( )**, which is used to set the entries in the nodal admittance matrix for a specific frequency value.

```
/************************************************/
/*                                              */
/*   Listing 9.2                                */
/*                                              */
/*   BuildAugNodalEqn()                         */
/*                                              */
/************************************************/

#include "globDefs.h"
#include "protos.h"
extern FILE *inFile;
extern FILE *outFile;
static fpos_t *inPtr;

int BuildAugNodalEqn( struct complex T[][MAX_COLUMNS],
                      struct complex W[],
                      int *matrixSize,
                      char dependentVariable[][5])
{
enum componentTypes component;
char depVar[5], depVar2[5];
char labels[5][10];
real values[5];
int hNode, iNode, jNode, kNode;
int maxNode, newN;
int i,j;
int parameterCount;
logical done;
```

```
if(fgetpos(inFile, inPtr) != 0) fprintf(outFile,"fgetpos failure\n");

/*-----------------------------------------*/
/*  zero out the T matrix and W vector    */
for(i=0; i<MAX_COLUMNS; i++)  W[i] = cmplx(0.0,0.0);
for(i=0; i<MAX_COLUMNS; i++)
    {
    for(j=0; j<MAX_COLUMNS; j++) T[i][j] = cmplx(0.0,0.0);
    }
maxNode = 0;
done = FALSE;


/*----------------------------------------------------------------*/
/* first pass just picks up components having admittance description */
for(;;)
    {
    GetBranchData( &component, labels, values, &hNode, &iNode,
                    &jNode, &kNode, &done, depVar, depVar2);

    if( jNode > maxNode ) maxNode = jNode;
    if( kNode > maxNode ) maxNode = kNode;
    if( hNode > maxNode ) maxNode = hNode;
    if( iNode > maxNode ) maxNode = iNode;

    switch (component)
        {
        case conductance:           /*  see Rule 9.1  */
            T[jNode][jNode].Re += values[0];
            T[kNode][kNode].Re += values[0];
            T[jNode][kNode].Re -= values[0];
            T[kNode][jNode].Re -= values[0];
            break;

        case capacitor:                 /*  see Rule 9.3  */
            T[jNode][jNode].Im += values[0];
            T[kNode][kNode].Im += values[0];
            T[jNode][kNode].Im -= values[0];
            T[kNode][jNode].Im -= values[0];
            break;

        case currentSource:             /*  see Rule 9.5  */
            W[jNode].Re -= values[0];
            W[kNode].Re += values[0];
            break;
```

```
case VCCS:          /*  see Rule 9.6  */
    T[jNode][hNode].Re += values[0];
    T[kNode][iNode].Re += values[0];
    T[jNode][iNode].Re -= values[0];
    T[kNode][hNode].Re -= values[0];
    break;
case resistor:
case inductor:
case voltageSource:
case VCVS:
case CCCS:
case CCVS:
case transformer:
case OpAmp:
    break;
        default:
            fprintf(outFile,"invalid component in pass 1\n");
            return -1;
        }
    if(done) break;
    }
/*-------------------------------------------*/
/*  write the names of the node voltages to  */
/*  a string array for later use             */
for(i=0; i<=maxNode; i++)
    {
    sprintf( &(dependentVariable[i][0]),"x%d\0",i);
    }
newN = maxNode;
/*-----------------------------------------------*/
/*  second pass to augment the basic Y partition   */

if(fsetpos(inFile, inPtr) != 0) fprintf(outFile,"fsetpos failure\n");
done = FALSE;

for(;;)
    {
    GetBranchData(   &component, labels, values, &hNode, &iNode,
                     &jNode, &kNode, &done, depVar, depVar2);

switch (component)
    {
    case resistor:                  /*  see Rule 9.7  */
        newN++;
        T[jNode][newN].Re = 1.0;
```

```
        T[kNode][newN].Re = -1.0;
        T[newN][newN].Re = -values[0];
        T[newN][jNode].Re = 1.0;
        T[newN][kNode].Re = -1.0;
        sprintf( &(dependentVariable[newN][0]),"%s",depVar);
        break;
    case inductor:                  /*  see Rule 9.7  */
        newN++;
        T[jNode][newN].Re = 1.0;
        T[kNode][newN].Re = -1.0;
        T[newN][newN].Im = -values[0];
        T[newN][jNode].Re = 1.0;
        T[newN][kNode].Re = -1.0;
        sprintf( &(dependentVariable[newN][0]),"%s",depVar);
        break;

    case voltageSource:             /*  see Rule 9.8  */
        newN++;
        T[jNode][newN].Re = 1.0;
        T[kNode][newN].Re = -1.0;
        T[newN][jNode].Re = 1.0;
        T[newN][kNode].Re = -1.0;
        W[newN].Re = values[0];
        sprintf( &(dependentVariable[newN][0]),"%s",depVar);
        break;
    case VCVS:                      /*  see Rule 9.9  */
        newN++;
        T[jNode][newN].Re = 1.0;
        T[kNode][newN].Re = -1.0;
        sprintf( &(dependentVariable[newN][0]),"%s",depVar);
        T[newN][hNode].Re = -values[0];
        T[newN][iNode].Re = values[0];
        T[newN][jNode].Re = 1.0;
        T[newN][kNode].Re = -1.0;
        break;

    case CCCS:                      /*  see Rule 9.10  */
        newN++;
        T[hNode][newN].Re = 1.0;
        T[iNode][newN].Re = -1.0;
        T[jNode][newN].Re = values[0];
        T[kNode][newN].Re = -values[0];
        sprintf( &(dependentVariable[newN][0]),"%s",depVar);
        break;
```

```
case CCVS:                    /*  see Rule 9.11  */
     newN++;
     T[newN][hNode].Re = 1.0;
     T[newN][iNode].Re = -1.0;
     T[hNode][newN].Re = 1.0;
     T[iNode][newN].Re = -1.0;
     T[newN+1][newN].Re = -values[0];
     newN++;
     T[newN][jNode].Re = 1.0;
     T[newN][kNode].Re = -1.0;
     T[jNode][newN].Re = 1.0;
     T[kNode][newN].Re = -1.0;
     sprintf( &(dependentVariable[newN][0]),"%s",depVar);
     break;

case transformer:          /*  see Rule 9.12  */
     newN++;
     T[newN][hNode].Re = 1.0;
     T[newN][iNode].Re = -1.0;
     T[newN][newN].Im = -values[0];
     T[hNode][newN].Re = 1.0;
     T[iNode][newN].Re = -1.0;
     T[newN+1][newN].Im = -values[2];
     T[newN][newN+1].Im = -values[2];
     newN++;
     T[newN][jNode].Re = 1.0;
     T[newN][kNode].Re = -1.0;
     T[newN][newN].Im = -values[1];

     T[jNode][newN].Re = 1.0;
     T[kNode][newN].Re = -1.0;
     sprintf( &(dependentVariable[newN][0]),"%s",depVar);
     sprintf( &(dependentVariable[newN][0]),"%s",depVar2);
     break;

case OpAmp:                   /*  see Rule 9.13  */
     newN++;
     T[newN][hNode].Re = 1.0;
     T[newN][iNode].Re = -1.0;
     T[jNode][newN].Re = 1.0;
     T[kNode][newN].Re = -1.0;
     break;

default:
     break;
```

```
        }
    if(done) break;
    }
*matrixSize = newN;
return 0;
}


/*********************************************/
/*                                           */
/*   Listing 9.3                             */
/*                                           */
/*   SetFreq()                               */
/*                                           */
/*********************************************/
#include "globDefs.h"
#include "protos.h"

int SetFreq( real freq,
             struct complex Y_of_s[][MAX_COLUMNS],
             int N,
             struct complex Y[][MAX_COLUMNS])
{
real omega;
int i, j;
omega = 2.0 * PI * freq;

for(i=0; i<=N; i++)
    {
    for(j=0; j<=N; j++)
        {
        Y[i][j].Im = Y_of_s[i][j].Im * omega;
        Y[i][j].Re = Y_of_s[i][j].Re;
        }
    }
return 0;
}
```

Chapter

10

Synthesis of Passive Networks

This chapter begins with a discussion of how specified input impedances can be realized as *LC* or *RC* networks. Several standard techniques for the synthesis of one-port networks are presented. The design of a one-port exhibiting a given input impedance is rarely a goal in and of itself. However, the techniques for synthesizing such one-ports are very important for two reasons: One-ports are often used as building blocks in the design of larger circuits, and as we will show in later sections, one-port synthesis techniques play a role in the synthesis of two-port networks.

10.1 Input Impedance

A linear one-port network is described by either its input impedance $Z(s)$ or its input admittance $Y(s)$, which are defined as

$$Z(s) = \frac{V(s)}{I(s)} = \frac{1}{Y(s)}$$

where $V(s)$ and $I(s)$ are the input voltage and input current.

Consider an impedance function $Z(s)$ of the form $Z(s) = P(s)/Q(s)$ where $P(s)$ and $Q(s)$ are polynomials in s. For $Z(s)$ to be the input impedance of an *LC* network, Rule 10.1 must be satisfied.

Rule 10.1 (Realizability conditions for *LC* impedances). A given function $Z(s)$ is realizable as the input impedance of an *LC* one-port provided that the following conditions are satisfied:

1. There is either a simple pole or a simple zero of $Z(s)$ at the origin of the s plane.
2. All poles and zeros of $Z(s)$ are located on the imaginary axis.

3. All poles and zeros of $Z(s)$ are simple, and except for a pole or zero at the origin, they occur in complex conjugate pairs.
4. Poles and zeros of $Z(s)$ alternate along the imaginary axis. There are never two poles without an intervening zero, and there are never two zeros without an intervening pole.

As a consequence of Rule 10.1, any $Z(s)$ that is realizable as an *LC* one-port will have an odd order, and the order of the numerator exceeds that of the denominator or vice versa.

An impedance that cannot be realized by an *LC* network might possibly be realizable as an *RC* network. For an impedance function $Z(s)$ to be realizable as the input impedance of an *RC* network, Rule 10.2 must be satisfied.

Rule 10.2 (Realizability conditions for *RC* impedances). A given function $Z(s)$ is realizable as the input impedance of an *RC* one-port provided that the following conditions are satisfied:

1. All poles and zeros of $Z(s)$ are located on the nonpositive real axis.
2. One pole of $Z(s)$ is either at the origin or closer to the origin than every zero of $Z(s)$.
3. Poles and zeros of $Z(s)$ alternate along the negative real axis.

Impedance scaling

To change the input impedance of a network from $Z(s)$ to $kZ(s)$, multiply each resistor and each inductor by k and divide each capacitor by k. (It is assumed that $k > 0$.) This process is called *magnitude scaling*. Any linear network containing inductors or capacitors will have an input impedance that is, in general, a function of frequency. To *frequency-scale* the impedance from $Z(s)$ to $Z(ks)$, multiply each capacitance and each inductance by k. Frequency scaling of response data was discussed in Chap. 5.

10.2 Foster Synthesis

Foster synthesis uses the partial fraction expansion of a given immittance function to obtain element values for a circuit that realizes the given immittance. Specifically, a Foster I network is based upon a partial fraction expansion of the input impedance $Z(s)$, and a Foster II network is based upon a partial fraction expansion of the input admittance $Y(s)$.

Foster I *LC* networks

A Foster I network has the form shown in Fig. 10.1. The series inductor and series capacitor enclosed by the dashed line may or may not be present depending upon the form of $Z(s)$. The inductor corresponds to an impedance

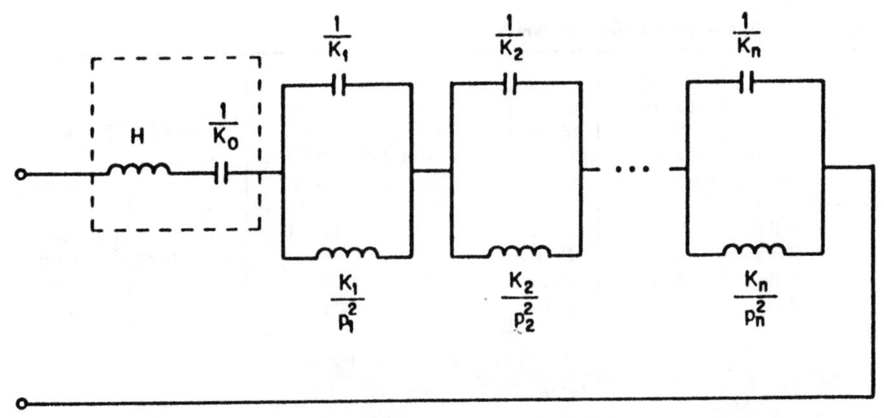

Figure 10.1 Circuit configuration for Foster I realization. (The series elements enclosed within the dashed line may or may not be present, in accordance with Table 10.1.)

pole at infinity, and the capacitor corresponds to an impedance pole at the origin. The parallel *LC* combinations correspond to conjugate poles on the imaginary axis. When the order of the denominator in $Z(s)$ exceeds the order of the numerator, both series elements are omitted for the case of a zero at the origin, and the series capacitor alone is used for the case of a pole at the origin. When the order of the numerator in $Z(s)$ exceeds the order of the denominator, both series elements are included for the case of a pole at the origin, and the series inductor alone is used for the case of a zero at the origin. This information is summarized in Table 10.1. Element values for a Foster I realization are obtained from a partial fraction expansion of $Z(s)$.

Algorithm 10.1 (Foster I *LC* synthesis). This algorithm synthesizes a network of the form shown in Fig. 10.1 to realize a given impedance $Z(s)$ provided that $Z(s)$ satisfies Rule 10.1.

Step 1. Determine the poles and zeros of the impedance function $Z(s)$. If the numerator and denominator are not in factored form, the root-finding techniques of Sec. 5.4 may prove useful in the factorization process.

Step 2. Examine the poles and zeros to determine whether they satisfy the conditions of Rule 10.1 and consequently whether $Z(s)$ is realizable as the input impedance of an *LC* network.

Step 3. Match the form of $Z(s)$ to one of the four cases in Table 10.1. Table entries indicate which of the optional series elements will be present in the Foster I realization of $Z(s)$. The circuit schematic can now be drawn.

Step 4. Use Algorithm 5.2 to obtain the partial fraction expansion of $Z(s)$. The result will be in the form

$$Z(s) = Hs + \frac{K_0}{s} + \frac{K_1 s}{s^2 + p_1^2} + \frac{K_2 s}{s^2 + p_2^2} + \cdots + \frac{K_n s}{s^2 + p_n^2}$$

TABLE 10.1 Properties of Foster Realizations

Equation	$Z(s)$ feature at origin	Features greatest finite distance from origin	$Z(s)$ feature at ∞	$Z(s)$ numerator order	$Z(s)$ denominator order	Foster I series element	Foster II shunt element
(1)	Pole	Poles	Zero	$2n$	$2n+1$	Capacitor	Capacitor
(2)	Zero	Zeros	Pole	$2n+1$	$2n$	Inductor	Inductor
(3)	Pole	Zeros	Pole	$2n+2$	$2n+1$	Both	Neither
(4)	Zero	Poles	Zero	$2n-1$	$2n$	Neither	Both

$$Z(s) = H\frac{(s^2+z_1^2)(s^2+z_2^2)\cdots(s^2+z_n^2)}{s(s^2+p_1^2)(s^2+p_2^2)\cdots(s^2+p_n^2)} \tag{1}$$

$$Z(s) = H\frac{s(s^2+z_1^2)(s^2+z_2^2)\cdots(s^2+z_n^2)}{(s^2+p_1^2)(s^2+p_2^2)\cdots(s^2+p_n^2)} \tag{2}$$

$$Z(s) = H\frac{(s^2+z_1^2)(s^2+z_2^2)\cdots(s^2+z_{n+1}^2)}{s(s^2+p_1^2)(s^2+p_2^2)\cdots(s^2+p_n^2)} \tag{3}$$

$$Z(s) = H\frac{s(s^2+z_1^2)(s^2+z_2^2)\cdots(s^2+z_{n-1}^2)}{(s^2+p_1^2)(s^2+p_2^2)\cdots(s^2+p_n^2)} \tag{4}$$

where p_i is obtained from the ith conjugate pair of pole frequencies

$$\omega_{pi} = \pm jp_i$$

Step 5. Obtain the circuit element values from $H, K_0, K_1, K_2, \ldots, K_n$ as follows:

$H =$ inductance of series inductor (if present)

$\dfrac{1}{K_0} =$ capacitance of series capacitor (if present)

$\dfrac{1}{K_n} =$ capacitance in nth parallel LC combination

$\dfrac{K_n}{p_n^2} =$ inductance in nth parallel LC combination ■

Example 10.1 Obtain the Foster I realization for

$$Z(s) = \frac{s^2+4}{s(s^2+25)} \tag{10.1}$$

solution Examination of (10.1) reveals that there are poles at $s = 0$ and $s = \pm 5j$, plus zeros at $s = \pm 2j$. This corresponds to the case represented by Eq. (1) of Table 10.1. Therefore,

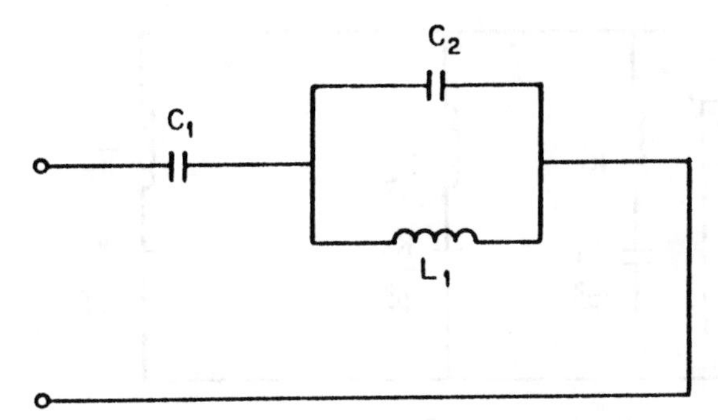

Figure 10.2 Foster I realization for Example 10.1.

the circuit will be as shown in Fig. 10.2. Partial fraction expansion of $Z(s)$ yields

$$Z(s) = \frac{K_0}{s} + \frac{K_1 s}{s(s^2+25)}$$

where

$$K_0 = \left.\frac{s^2+4}{s^2+25}\right|_{s=0} = \frac{4}{25}$$

$$K_1 = \left.\frac{s^2+4}{s^2}\right|_{s^2=-25} = \frac{21}{25}$$

Thus

$$C_1 = \frac{1}{K_0} = 6.25 \text{ F}$$

$$C_2 = \frac{1}{K_1} = 1.1905 \text{ F}$$

$$L_1 = \frac{K_1}{p_1^2} = 33.6 \text{ mH}$$

Foster II networks

A Foster II network has the form shown in Fig. 10.3. The shunt inductor and shunt capacitor enclosed by the dashed line may or may not be present depending upon the form of $Z(s)$. The inductor corresponds to an impedance zero at the origin, and the capacitor corresponds to an impedance zero at infinity. The series LC combinations correspond to conjugate zeros on the imaginary axis. When the order of the denominator in $Z(s)$ exceeds the order of the numerator, both shunt elements are included for the case of a zero at the origin and the shunt capacitor alone is used for the case of a pole at the origin. When the order of the numerator in $Z(s)$ exceeds the order of the denominator, both shunt elements are omitted for the case of a pole at the origin and the shunt inductor alone is used for the case of a zero at the origin.

This information is summarized in Table 10.1. Element values for a Foster II realization are obtained from a partial fraction expansion of the admittance function $Y(s)$.

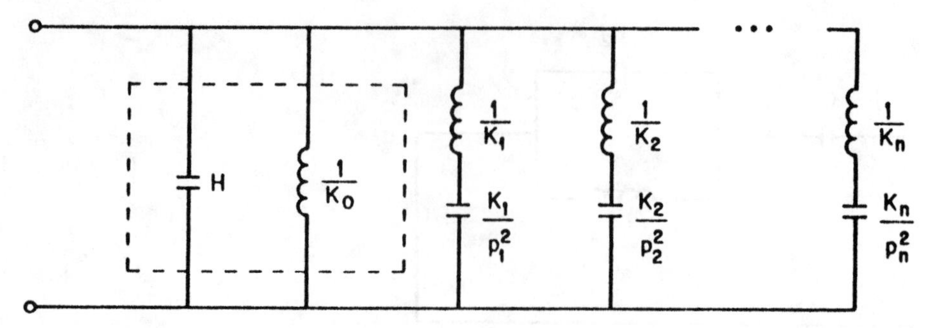

Figure 10.3 Circuit configuration for Foster II realization. (The shunt elements enclosed within the dashed line may or may not be present, in accordance with Table 10.1.)

Algorithm 10.2 (Foster II synthesis). This algorithm synthesizes a network of the form shown in Fig. 10.3 to realize a given impedance $Z(s)$ provided that $Z(s)$ satisfies Rule 10.1.

Step 1. Determine the poles and zeros of the impedance function $Z(s)$. If the numerator and denominator are not in factored form, the root-finding techniques of Sec. 5.4 may prove useful in the factorization process.

Step 2. Examine the poles and zeros to determine whether they satisfy the conditions of Rule 10.1 and, consequently, whether $Z(s)$ is realizable as the input impedance of an LC network. If Rule 10.1 is not satisfied, none of the Foster or Cauer synthesis algorithms can be used.

Step 3. Match the form of $Z(s)$ to one of the four cases in Table 10.1. Table entries indicate which of the optional shunt elements will be present in the Foster II realization of $Z(s)$. The circuit schematic can now be drawn.

Step 4. Use the techniques of Sec. 5.5 to obtain the partial fraction expansion of $Y(s) = 1/Z(s)$. The result will be in the form

$$Y(s) = Hs + \frac{K_0}{s} + \frac{K_1 s}{s^2 + p_1^2} + \frac{K_2 s}{s^2 + p_2^2} + \cdots + \frac{K_n s}{s^2 + p_n^2}$$

where p_i is obtained from the ith conjugate pair of admittance pole frequencies $\omega_{pi} = \pm jp_i$.

Step 5. Obtain the circuit element values from $H, K_0, K_1, K_2, \ldots, K_n$ as follows:

$H =$ capacitance of shunt capacitor (if present)

$\dfrac{1}{K_0} =$ inductance of shunt inductor (if present)

$\dfrac{1}{K_n} =$ inductance in nth series LC combination

$\dfrac{K_n}{p_n^2} =$ capacitance in nth series LC combination ∎

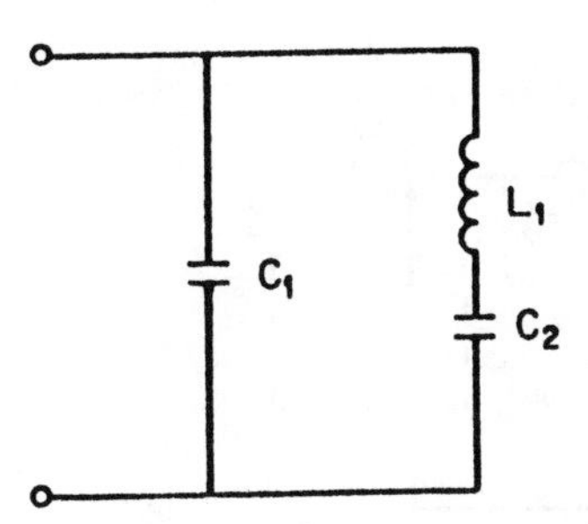

Figure 10.4 Foster II realization for Example 10.2.

Example 10.2 Obtain the Foster II realization for

$$Y(s) = \frac{s(s^2 + 25)}{s^2 + 4} \tag{10.2}$$

solution Examination of (10.2) reveals that $Z(s) = 1/Y(s)$ has poles at $s = 0$ and $s = \pm 5j$ plus zeros at $s = \pm 2j$. This corresponds to the case represented by Eq. (1) in Table 10.1. Therefore the circuit will have a shunt capacitor but no shunt inductor, as shown in Fig. 10.4. The order of the numerator in (10.2) exceeds the order of the denominator, so the first step of partial fraction expansion entails division, to obtain

$$Y(s) = s + \frac{21s}{s^2 + 4} \tag{10.3}$$

The second term in (10.3) needs no further expansion, so we conclude that $H = 1$ and $K_1 = 21$ and thus

$$C_1 = 1\text{ F}$$

$$C_2 = {}^{21}\!/_{4} = 5.25\text{ F}$$

$$L = {}^{1}\!/_{21} = 47.62\text{ mH}$$

10.3 Cauer Synthesis

Cauer synthesis uses the continued-fraction expansion of a given immittance function to obtain element values for a circuit that realizes the given immittance. Cauer I networks contain series inductors and shunt capacitors. Cauer II networks contain series capacitors and shunt inductors.

Cauer I networks

A Cauer I network has the form shown in Fig. 10.5. The elements enclosed by the dashed line may or may not be present depending upon the form of $Z(s)$. The inductor corresponds to an impedance pole at infinity, and the capacitor corresponds to an admittance zero (and hence an impedance pole) at the origin.

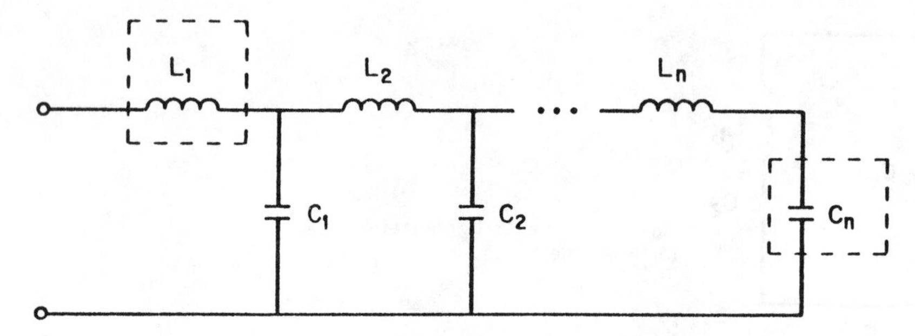

Figure 10.5 Circuit configuration for Cauer I realization. (The elements enclosed within the dashed lines may or may not be present, in accordance with Table 10.2.)

Algorithm 10.3 (Cauer I synthesis). This algorithm synthesizes a network of the form shown in Fig. 10.5 to realize a given impedance $Z(s)$ provided that $Z(s)$ satisfies Rule 10.1.

Step 1. Determine the poles and zeros of the impedance function $Z(s)$. If the numerator and denominator are not in factored form, the root-finding techniques of Sec. 5.4 may prove useful in the factorization process.

Step 2. Examine the poles and zeros to determine whether they satisfy the conditions of Rule 10.1 and, consequently, whether $Z(s)$ is realizable as the input impedance of an LC network.

Step 3. Match the form of $Z(s)$ to one of the four cases in Table 10.2. Table entries indicate which of the optional elements will be present in the Cauer

TABLE 10.2 Properties of Cauer Realizations

Equation	$Z(s)$ feature at origin	Features greatest finite distance from origin	$Z(s)$ feature at ∞	Cauer I optional elements	Cauer I function to expand	Cauer II optional elements	Cauer II function to expand
(1)	Pole	Poles	Zero	Capacitor	$Y(s)$	Capacitor	$Z(s)$
(2)	Zero	Zeros	Pole	Inductor	$Z(s)$	Inductor	$Y(s)$
(3)	Pole	Zeros	Pole	Both	$Z(s)$	Both	$Z(s)$
(4)	Zero	Poles	Zero	Neither	$Y(s)$	Neither	$Y(s)$

$$Z(s) = H\frac{(s^2+z_1^2)(s^2+z_2^2)\cdots(s^2+z_n^2)}{s(s^2+p_1^2)(s^2+p_2^2)\cdots(s^2+p_n^2)} \tag{1}$$

$$Z(s) = H\frac{s(s^2+z_1^2)(s^2+z_2^2)\cdots(s^2+z_n^2)}{(s^2+p_1^2)(s^2+p_2^2)\cdots(s^2+p_n^2)} \tag{2}$$

$$Z(s) = H\frac{(s^2+z_1^2)(s^2+z_2^2)\cdots(s^2+z_{n+1}^2)}{s(s^2+p_1^2)(s^2+p_2^2)\cdots(s^2+p_n^2)} \tag{3}$$

$$Z(s) = H\frac{s(s^2+z_1^2)(s^2+z_2^2)\cdots(s^2+z_{n-1}^2)}{(s^2+p_1^2)(s^2+p_2^2)\cdots(s^2+p_n^2)} \tag{4}$$

I realization of $Z(s)$. The number of elements in the circuit equals the number of poles in $Z(s)$ (including a pole at infinity if one is indicated by the table). The circuit schematic can now be drawn.

Step 4. Arrange, into descending powers of s, the unfactored numerator and denominator polynomials; then use the techniques of Sec. 5.6 to compute the continued-fraction expansion of either $Y(s)$ or $Z(s)$, as indicated by Table 10.2. The result will have the form

$$F(s) = b_0 s + \cfrac{1}{b_1 s + \cfrac{1}{b_2 s + \cfrac{1}{b_3 s + \cdots}}}$$

Step 5. Obtain the inductor values from the continued-fraction partial denominators b_k. If the optional series inductor L_1 is present, the inductor values are given by

$$L_k = b_{2k-2} \qquad k = 1, 2, \ldots, n$$

If the optional inductor is omitted, the inductor values are given by

$$L_k = b_{2k-3} \qquad k = 2, 3, \ldots, n$$

Step 6. Obtain the capacitor values from the continued-fraction partial denominators b_k. If the optional series inductor L_1 is present, the capacitor values are given by

$$C_k = b_{2k-1} \qquad k = 1, 2, \ldots, m$$

If the optional inductor is omitted, the capacitor values are given by

$$C_k = b_{2k-2} \qquad k = 1, 2, \ldots, m$$

The final index value m equals either n or $n - 1$ depending upon whether the optional shunt capacitor is included. ■

Example 10.3 Obtain the Cauer I realization for

$$Z(s) = \frac{s^4 + 10s^2 + 9}{s^3 + 4s}$$

solution The numerator and denominator of $Z(s)$ can be factored to obtain

$$Z(s) = \frac{(s^2 + 1)(s^2 + 9)}{s(s^2 + 4)}$$

Thus $Z(s)$ has poles at $s = 0$ and $s = \pm 2j$ plus zeros at $s = \pm j$ and $s = \pm 3j$. Rule 10.1 is satisfied, so we can proceed. The pole-zero pattern corresponds to the case represented by Eq. (3) of Table 10.2. The table indicates that both optional elements are included, and there is a pole at infinity. Therefore, the circuit will contain four elements configured as shown in Fig. 10.6. Arranging the numerator and denominator of $Z(s)$ into descending

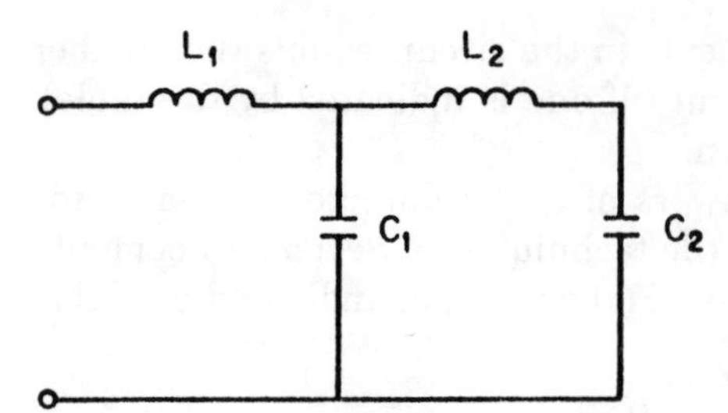

Figure 10.6 Cauer I realization for Example 10.3.

powers of s and performing the continued-fraction expansion result in

$$Z(s) = s + \cfrac{1}{\frac{1}{6}s + \cfrac{1}{\frac{12}{5}s + \cfrac{1}{\frac{5}{18}s}}}$$

Therefore, we conclude that

$$L_1 = 1 \qquad L_2 = 12/5 \qquad C_1 = 1/6 \qquad C_2 = 5/18$$

Example 10.4 Obtain the Cauer I realization for

$$Z(s) = \frac{s^4 + 10s^2 + 9}{s^5 + 20s^3 + 64s}$$

solution The numerator and denominator of $Z(s)$ can be factored to obtain

$$Z(s) = \frac{(s^2+1)(s^2+9)}{s(s^2+4)(s^2+16)}$$

Thus $Z(s)$ has poles at $s = 0$, $s = \pm 2j$, and $s = \pm 4j$ plus zeros at $s = \pm j$ and $s = \pm 3j$. Rule 10.1 is satisfied, so we can proceed. The pole-zero pattern of $Z(s)$ corresponds to the case represented by Eq. (1) in Table 10.2. The table indicates that the optional inductor is omitted, so the circuit will contain five elements configured as shown in Fig. 10.7. Arranging the numerator and denominator of $Y(s)$ into descending powers of s and performing the continued-fraction expansion result in

$$Y(s) = s + \cfrac{1}{\frac{1}{10}s + \cfrac{1}{\frac{20}{9}s + \cfrac{1}{\frac{9}{70}s + \cfrac{1}{\frac{35}{9}s}}}}$$

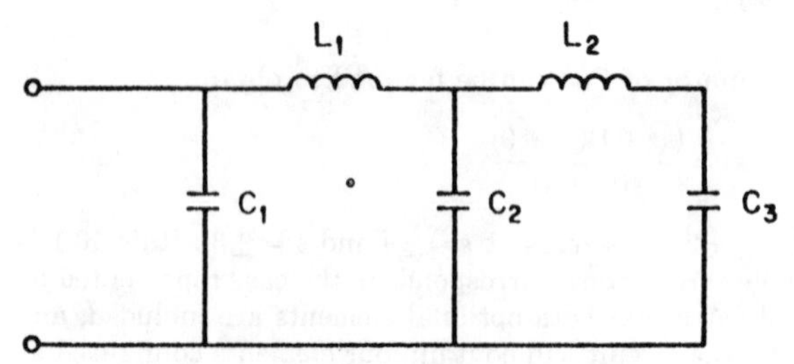

Figure 10.7 Cauer I realization for Example 10.4.

Thus $L_2 = b_1 = 1/10 \qquad L_3 = b_3 = 9/70$

$$C_1 = b_0 = 1 \qquad C_2 = b_2 = 20/9 \qquad C_3 = b_4 = 35/9$$

Cauer II networks

A Cauer II network has the form shown in Fig. 10.8. The elements enclosed by the dashed line may or may not be present depending upon the form of $Z(s)$. The capacitor corresponds to an impedance pole at the origin, and the inductor corresponds to an impedance pole at infinity.

Algorithm 10.4 (Cauer II synthesis). This algorithm synthesizes a network of the form shown in Fig. 10.8 to realize a given impedance $Z(s)$ provided that $Z(s)$ satisfies Rule 10.1.

Step 1. Determine the poles and zeros of the impedance function $Z(s)$. If the numerator and denominator are not in factored form, the root-finding techniques of Sec. 5.4 may prove useful in the factorization process.

Step 2. Examine the poles and zeros to determine whether they satisfy the conditions of Rule 10.1 and, consequently, whether $Z(s)$ is realizable as the input impedance of an LC network.

Step 3. Match the form of $Z(s)$ to one of the four cases in Table 10.2. Table entries indicate which of the optional elements will be present in the Cauer II realization of $Z(s)$. The number of elements in the circuit equals the number of poles in $Z(s)$ (including a pole at infinity if one is indicated by the table). The circuit schematic can now be drawn.

Step 4. Arrange, into ascending powers of s, the unfactored numerator and denominator polynomials; then use the techniques of Sec. 5.6 to compute the continued-fraction expansion of either $Y(s)$ or $Z(s)$, as indicated by Table 10.2. The result will have the form

$$F(s) = \frac{b_0}{s} + \cfrac{1}{\cfrac{b_1}{s} + \cfrac{1}{\cfrac{b_2}{s} + \cfrac{1}{\cfrac{b_3}{s} + \cdots}}}$$

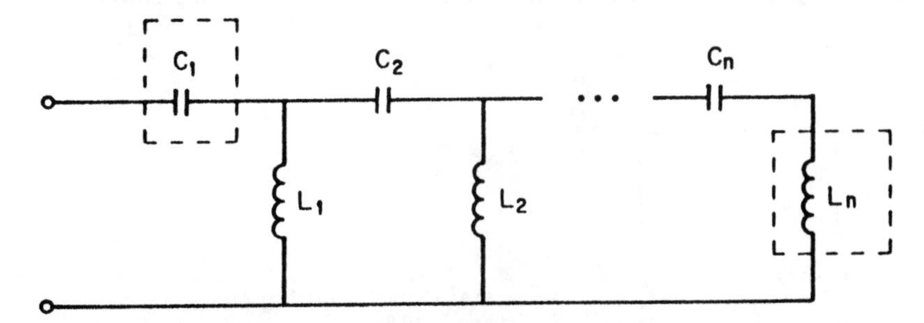

Figure 10.8 Circuit configuration for Cauer II realization. (The elements enclosed within the dashed lines may or may not be present, in accordance with Table 10.2.)

Step 5. Obtain the inductor values from the continued-fraction partial denominators b_k. If the optional series capacitor C_1 is present, the capacitor values are given by

$$C_k = \frac{1}{b_{2k-2}} \qquad k = 1, 2, \ldots, n$$

If the optional capacitor is omitted, the capacitor values are given by

$$C_k = \frac{1}{b_{2k-3}} \qquad k = 2, 3, \ldots, n$$

Step 6. Obtain the inductor values from the continued-fraction partial denominators b_k. If the optional series capacitor C_1 is present, the inductor values are given by

$$L_k = \frac{1}{b_{2k-1}} \qquad k = 1, 2, \ldots, m$$

If the optional capacitor is omitted, the inductor values are given by

$$L_k = \frac{1}{b_{2k-2}} \qquad k = 1, 2, \ldots, m$$

The final index value m equals either n or $n-1$ depending upon whether the optional shunt inductor is included. ■

Example 10.5 Obtain the Cauer II realization for

$$Z(s) = \frac{s^4 + 10s^2 + 9}{s^3 + 4s} = \frac{(s^2+1)(s^2+9)}{s(s^2+4)}$$

solution Rule 10.1 is satisfied since $Z(s)$ has poles at $s = 0$ and $s = \pm 2j$ plus zeros at $s = \pm j$ and $s = \pm 3j$. The pole-zero pattern corresponds to the case represented by Eq. (3) of Table 10.2. The table indicates that both optional elements are included, and there is a pole at infinity. Therefore, the circuit will contain four elements configured as shown in Fig. 10.9. Arranging the numerator and denominator of $Z(s)$ into ascending powers of s

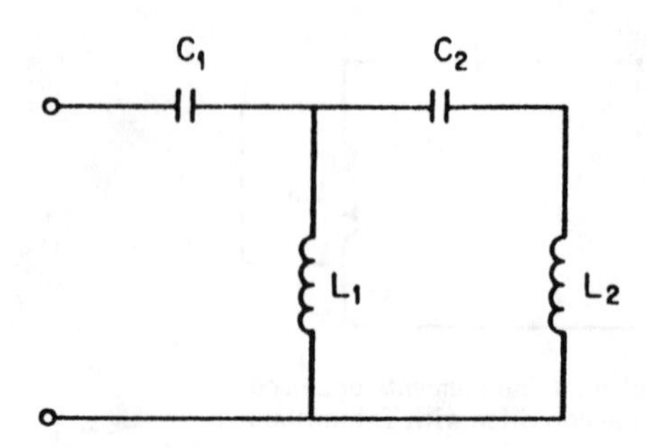

Figure 10.9 Cauer II realization for Example 10.5.

and computing the continued-fraction expansion result in

$$Z(s) = \frac{9}{4s} + \cfrac{1}{\cfrac{16}{31s} + \cfrac{1}{\cfrac{961}{60s} + \cfrac{1}{\cfrac{31}{15s}}}}$$

Thus $b_0 = 9/4 \quad b_1 = 16/31 \quad b_2 = 961/60 \quad b_3 = 31/15$

$C_1 = 4/9 \quad C_2 = 60/961 \quad L_1 = 31/16 \quad L_2 = 15/31$

10.4 Source-Terminated *LC* Ladder Networks

The poles of a lossless *LC* network all lie on the imaginary axis, but adding a resistance anywhere in the network will move the poles off the axis and into the left half-plane. A resistor can be located at the input as shown in Fig. 10.10. For such a configuration, it is possible to synthesize the *LC* subnetwork such that a specified voltage transfer V_0/V_{in} is obtained. This voltage transfer will have poles in the left half of the s plane. Let T_{LC} represent the voltage transfer of the *LC* network. Thus,

$$V_0 = V_2 = T_{LC} V_1$$
$$V_1 = \frac{V_0}{T_{LC}} \tag{10.4}$$

Let Z_{LC} denote the input impedance of the *LC* network. The combination of the resistor and the input impedance of the *LC* ladder can be viewed as a voltage divider, and therefore the voltage at node 1 can be expressed in terms of V_{in} as

$$V_1 = \frac{Z_{LC}}{1 + Z_{LC}} V_{in} \tag{10.5}$$

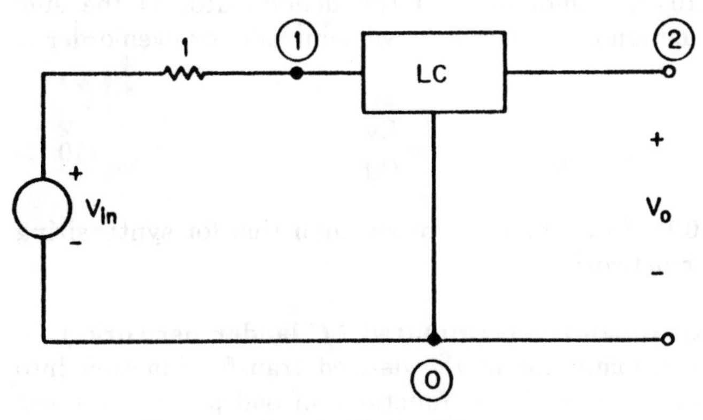

Figure 10.10 Source-terminated *LC* network.

Substitution of (10.4) into (10.5) and solving for V_0/V_{in} yield

$$\frac{V_0}{V_{\text{in}}}=\frac{Z_{LC}T_{LC}}{1+Z_{LC}} \tag{10.6}$$

The input impedance Z_{LC} will always be an odd-order function of s, and T_{LC} will always be an even-order function of s. Therefore, the numerator of (10.6) will always be an odd-order function of s. Setting this result aside for a moment, let us now assume that the denominator of V_0/V_{in} is separated into odd-order and even-order parts such that

$$\frac{V_0}{V_{\text{in}}}=\frac{N}{\text{Ev}+\text{Od}} \tag{10.7}$$

To make the form of (10.7) similar to the form of (10.6), we divide the numerator and denominator of (10.7) by either Od or Ev. If N is odd-order, we divide by Ev to obtain

$$\frac{V_0}{V_{\text{in}}}=\frac{N/\text{Ev}}{1+\text{Od}/\text{Ev}} \tag{10.8}$$

The numerators of both (10.6) and (10.8) are odd-order functions of s, and both denominators are sums of 1 and an odd-order function of s. Therefore, we conclude for odd-order N that

$$T_{LC}=\frac{N}{\text{Od}} \qquad Z_{LC}=\frac{\text{Od}}{\text{Ev}} \tag{10.9}$$

For even-order N we divide by Od to obtain

$$\frac{V_0}{V_{\text{in}}}=\frac{N/\text{Od}}{1+\text{Ev}/\text{Od}} \tag{10.10}$$

The numerator of (10.10) is odd-order, and the denominator is the sum of 1 and an odd-order function of s. Therefore, we conclude for even-order N that

$$T_{LC}=\frac{N}{\text{Ev}} \qquad Z_{LC}=\frac{\text{Ev}}{\text{Od}} \tag{10.11}$$

Equations (10.9) and (10.11) form the basis of an algorithm for synthesizing source-terminated ladder networks.

Algorithm 10.5 (Synthesis of source-terminated *LC* ladder networks)

Step 1. Separate the denominator of the desired transfer function into parts Od and Ev which are, respectively, functions of odd powers of s and even powers of s.

Step 2. If the numerator of the desired transfer function is an odd-order function of s, set $Z_{LC} = \text{Od}/\text{Ev}$. If the numerator of the transfer function is an even-order function of s, set $Z_{LC} = \text{Ev}/\text{Od}$.

Step 3. Use Algorithm 10.3 or 10.4 to synthesize an LC one-port having an input impedance of Z_{LC}.

Step 4. Determine the node in the LC ladder to which the output must be connected so that the correct number of zeros at the origin and at infinity are realized. (Shunt capacitors and series inductors produce zeros at infinity. Series capacitors and shunt inductors produce zeros at the origin.) ■

Example 10.6 Synthesize a source-terminated network that realizes the transfer function

$$H(s) = \frac{1}{s^5 + 3.236s^4 + 5.236s^3 + 5.236s^2 + 3.236s + 1} \tag{10.12}$$

solution The numerator of (10.12) is an even-order function of s, so we use Eq. (10.11) to obtain

$$Z_{LC}(s) = \frac{\text{Ev}}{\text{Od}} = \frac{3.236s^4 + 5.236s^2 + 1}{s^5 + 5.236s^3 + 3.236s}$$

We will use Algorithm 10.3 to obtain the Cauer I realization of Z_{LC}. The form of Z_{LC} corresponds to the case represented by Eq. (1) in Table 10.2. The table indicates that the optional inductor is omitted, so the circuit will contain five elements configured as shown in Fig. 10.11. The continued-fraction expansion of Y_{LC} is given by

$$Y_{LC}(s) = 0.3090s + \cfrac{1}{0.8944s + \cfrac{1}{1.382s + \cfrac{1}{1.695s + \cfrac{1}{1.545s}}}}$$

We therefore conclude that

$$C_1 = 0.3090 \text{ F} \qquad C_2 = 1.382 \text{ F} \qquad C_3 = 1.545 \text{ F}$$

$$L_1 = 894.4 \text{ mH} \qquad L_2 = 1695 \text{ mH}$$

Examination of (10.12) reveals that there are five zeros at infinity, so the input is connected as shown in Fig. 10.12 so that all the inductors are in series and all the capacitors are in shunt.

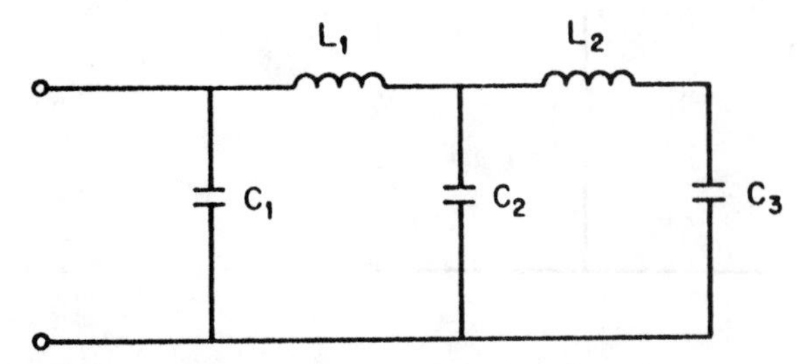

Figure 10.11 The LC one-port for Example 10.6.

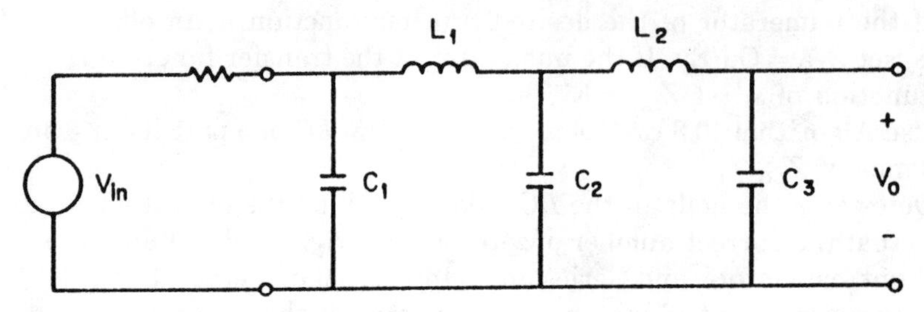

Figure 10.12 Completed network for Example 10.6.

10.5 Load-Terminated *LC* Ladder Networks

Section 10.4 showed how a resistor can be added to the input of a lossless LC ladder in order to move the poles off the imaginary axis. An alternative approach locates a resistor across the output as shown in Fig. 10.13. Again, it is possible to synthesize the LC subnetwork such that a specified voltage transfer V_0/V_{in} is obtained. However, the synthesis procedure differs slightly from the procedure for the source-terminated case. Let T_{LC} represent the voltage transfer, and let Z_{LCO} represent the output impedance of the unterminated LC ladder. The combination of the resistor and the output impedance of the LC ladder can be viewed as a voltage divider that divides the ladder's open-circuit output voltage $T_{LC}V_{\text{in}}$. Thus

$$V_0 = \frac{T_{LC}}{1 + Z_{LCO}} V_{\text{in}}$$

$$\frac{V_0}{V_{\text{in}}} = \frac{T_{LC}}{1 + Z_{LCO}} \tag{10.13}$$

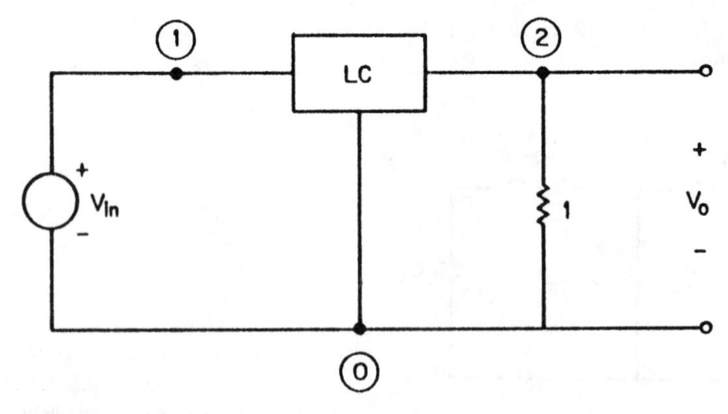

Figure 10.13 Load-terminated *LC* network.

The output impedance Z_{LCO} will always be an odd-order function of s, and T_{LC} will always be an even-order function of s. Assume that the denominator of V_0/V_{in} is separated into even and odd parts such that

$$\frac{V_0}{V_{in}} = \frac{N}{\text{Ev} + \text{Od}} \tag{10.14}$$

To make the form of (10.14) similar to the form of (10.13), we divide the numerator and denominator of (10.14) by either Od or Ev. If N is odd-order, we divide by Od to obtain

$$\frac{V_0}{V_{in}} = \frac{N/\text{Od}}{1 + \text{Ev}/\text{Od}} \tag{10.15}$$

The numerators of both (10.13) and (10.15) are even-order functions of s, and both denominators are sums of 1 and an odd-order function of s. Therefore, we conclude for odd-order N that

$$T_{LC} = \frac{N}{\text{Od}} \qquad Z_{LC} = \frac{\text{Ev}}{\text{Od}} \tag{10.16}$$

For even-order N we divide by Ev to obtain

$$\frac{V_0}{V_{in}} = \frac{N/\text{Ev}}{1 + \text{Od}/\text{Ev}} \tag{10.17}$$

The numerator of (10.17) is even-order, and the denominator is the sum of 1 and an odd-order function of s. Therefore, we conclude for even-order N that

$$T_{LC} = \frac{N}{\text{Ev}} \qquad Z_{LC} = \frac{\text{Od}}{\text{Ev}} \tag{10.18}$$

Equations (10.16) and (10.18) form the basis of an algorithm for synthesizing load-terminated LC ladder networks.

Algorithm 10.6 (Synthesis of load-terminated LC ladder networks)

Step 1. Separate the denominator of the desired transfer function into parts Od and Ev which are, respectively, functions of odd powers of s and even powers of s.

Step 2. If the numerator of the desired transfer function is an odd-order function of s, set $Z_{LCO} = \text{Ev}/\text{Od}$. If the numerator of the transfer function is an even-order function of s, set $Z_{LCO} = \text{Od}/\text{Ev}$.

Step 3. Use Algorithm 10.3 or 10.4 to synthesize an LC one-port having an input impedance of Z_{LCO}.

Step 4. Determine the node in the LC ladder to which the output must be connected so that the correct number of zeros at the origin and at infinity are realized. ■

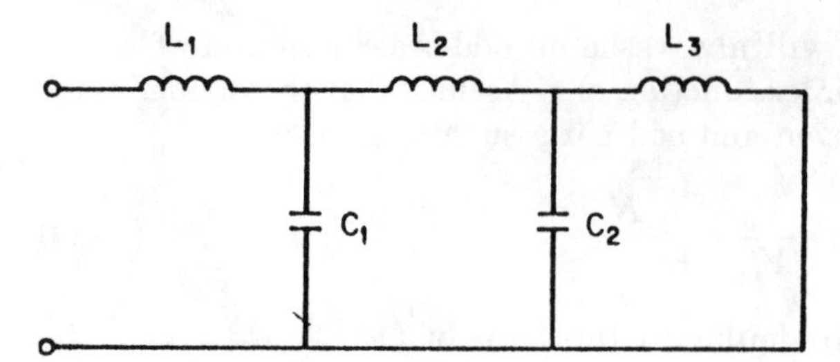

Figure 10.14 The *LC* one-port for Example 10.7.

Example 10.7 Synthesize a source-terminated network that realizes the transfer function

$$H(s) = \frac{1}{s^5 + 3.236s^4 + 5.236s^3 + 5.236s^2 + 3.236s + 1} \tag{10.19}$$

solution The numerator of (10.19) is an even-order function of s, so we use Eq. (10.18) to obtain

$$Z_{LCO} = \frac{\text{Od}}{\text{Ev}} = \frac{s^5 + 5.236s^3 + 3.236s}{3.236s^4 + 5.236s^2 + 1}$$

We will use Algorithm 10.3 to obtain the Cauer I realization of Z_{LCO}. The form of Z_{LCO} corresponds to the case represented by Eq. (2) in Table 10.2. The table indicates that the optional capacitor is omitted, so the circuit will contain five elements configured as shown in Fig. 10.14. The continued-fraction expansion of Z_{LCO} is given by

$$Z_{LCO} = 0.3090s + \cfrac{1}{0.8944s + \cfrac{1}{1.382s + \cfrac{1}{1.695s + \cfrac{1}{1.545s}}}}$$

We therefore conclude that

$$L_1 = 309 \text{ mH} \qquad L_2 = 1.382 \text{ H} \qquad L_3 = 1.545 \text{ H}$$

$$C_1 = 0.8944 \text{ F} \qquad C_2 = 1.695 \text{ F}$$

The *LC* ladder is redrawn in Fig. 10.15 to show the load termination. The desired transfer function has five zeros at infinity, so the input is connected as shown in Fig. 10.15 so that all the inductors are in series and all the capacitors are in shunt.

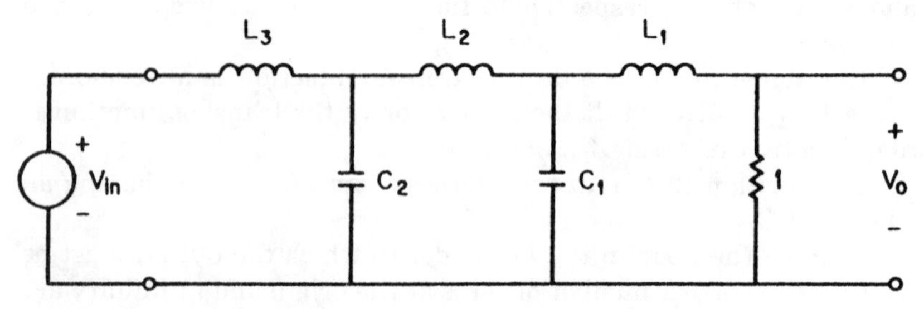

Figure 10.15 Completed network for Example 10.7.

10.6 Double-Terminated *LC* Ladder Networks

Previous sections have shown how transfer functions can be realized as lossless *LC* ladder networks with a resistive termination on either the input or the output. In practice, *LC* ladder networks are often terminated at both the input and the output, as shown in Fig. 10.16. This section presents a procedure for synthesis of double-terminated networks. The algorithm is based on a technique called *insertion-loss synthesis*.

Algorithm 10.7 (Synthesis of double-terminated *LC* ladder networks)

Step 1. Set

$$F(s) = |H(j\omega)|^2\big|_{\omega = s/j}$$

where $|H(j\omega)|^2$ is obtained either directly from the defining equations for a filter or from the transfer function $H(s)$ by using

$$|H(j\omega)|^2 = [H(s)H(-s)]\big|_{s=j\omega}$$

Step 2. Select the desired input resistance R_1 and output resistance R_2. The values for R_1 and R_2 must be chosen such that

$$\frac{4R_1R_2}{(R_1+R_2)^2}F(s) \leq 1 \tag{10.20}$$

For most lowpass response specifications normalized to have $|H(0)|^2 = 1$, Eq. (10.20) will usually imply that R_1 and R_2 must satisfy

$$\frac{4R_1R_2}{(R_1+R_2)^2} \leq 1$$

However, an even-order Chebyshev filter normalized to have $|H(0)|^2 = 1$ will exhibit ripples greater than 1. Therefore, the terminations for a normalized

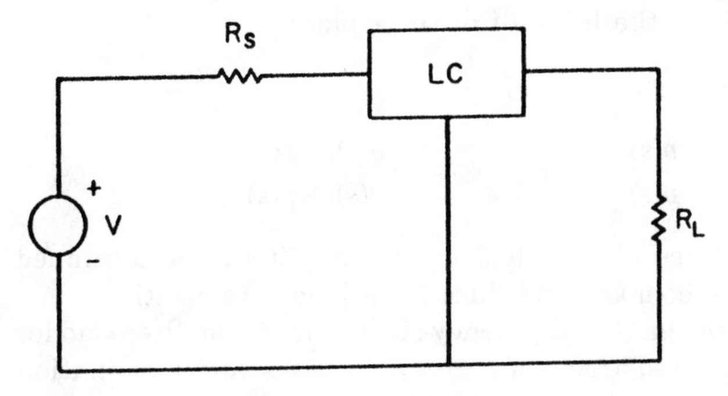

Figure 10.16 Double-terminated *LC* ladder.

even-order Chebyshev filter must satisfy

$$\frac{4R_1R_2}{(R_1+R_2)^2} \leq \frac{1}{1+\epsilon^2} \tag{10.21}$$

An implication of (10.21) is that an even-order Chebyshev filter cannot be realized with $R_1 = R_2$.

Step 3. From $F(s)$, R_1, and R_2 generate

$$\frac{A(s)}{B(s)} = 1 - \frac{4R_1R_2}{(R_1+R_2)^2} F(s)$$

such that $p(s)$ and $q(s)$ are polynomials in s.

Step 4. Solve for the zeros of $B(s)$. (The root-finding techniques of Sec. 5.4 may prove useful.)

Step 5. Form $q(s)$ from the zeros of $B(s)$ that lie in the left half of the s plane

$$q(s) = \prod_{\sigma_i < 0} (s - s_i)$$

where $s_i = \sigma_i + j\omega_i$.

Step 6. Solve for the zeros of $A(s)$.

Step 7. Form $p(s)$ from half of the zeros of $A(s)$

$$p(s) = \prod (s - s_i)$$

where the zeros s_i are selected according to the following rules:

- Complex zeros must be selected in conjugate pairs; i.e., if $a + bj$ is selected for $p(s)$, then $a - bj$ must also be selected.
- If a complex zero $s_i = a + bj$ is selected for $p(s)$, then the zero $-a + bj$ must not be selected.
- If a minimum-phase realization is desired, select zeros for $p(s)$ only from the zeros of $A(s)$ that lie in the left half of the s plane.

Step 8. Compute $Z(s)$ as

$$Z(s) = \frac{q(s) + p(s)}{q(s) - p(s)} \qquad \text{or} \qquad Z(s) = \frac{q(s) - p(s)}{q(s) + p(s)}$$

Step 9. Use the techniques of Sec. 10.3 to expand $Z(s)$ as a continued fraction, and determine the component values for a ladder realization.

Step 10. Impedance-scale and/or frequency-scale the normalized ladder network to obtain the required frequency characteristics and termination resistances. ■

Example 10.8 Use Algorithm 10.7 to synthesize a third-order Butterworth filter having a cutoff frequency of 15 kHz and termination resistances of $R_S = R_L = 1\ \text{k}\Omega$.

solution For a third-order Butterworth filter

$$F(s) = \frac{1}{1 - s^6}$$

Initially, we set the termination resistances each equal to 1 Ω. The normalized network will eventually be impedance-scaled to achieve the desired terminations of 1 kΩ. Thus we generate

$$\frac{A(s)}{B(s)} = 1 - \frac{(4)(1)(1)}{(1+1)^2} \cdot \frac{1}{1-s^6} = \frac{s^6}{s^6 - 1}$$

The zeros of $B(s)$ are found to be ± 1, $\pm \frac{1}{2} + (\sqrt{3}/2)j$, and $\pm \frac{1}{2} - (\sqrt{3}/2)j$. The left-half-plane zeros are -1 and $-\frac{1}{2} \pm (\sqrt{3}/2)j$. Thus,

$$q(s) = (s+1)\left(s + \frac{1}{2} + \frac{\sqrt{3}}{2}j\right)\left(s + \frac{1}{2} - \frac{\sqrt{3}}{2}j\right)$$

$$= s^3 + s^2 + 2s + 1$$

The numerator $A(s)$ has a zero of multiplicity 6 at $s = 0$; therefore, $p(s) = s^3$, and

$$Z(s) = \frac{2s^3 + 2s^2 + 2s + 1}{2s^2 + 2s + 1}$$

$$= s + \cfrac{1}{2s + \cfrac{1}{s + \cfrac{1}{1}}}$$

Therefore, the normalized network will be as shown in Fig. 10.17. To obtain $R_S = R_L = 1000\ \Omega$, the network must be impedance-scaled by a factor of 1000, thus yielding $L_1 = L_2 = 1000$ and $C = 2 \times 10^{-3}$. The frequency-scaling factor is $30{,}000\pi$, so the final component values are $L_1 = L_2 = 10.61$ mH and $C = 0.212\ \mu$F.

The third-order Butterworth filter is the most frequently given example of insertion-loss synthesis because the normalized component values turn out to be tidy little integers. In general, as we will see in Chap. 13, the numbers are usually much less cooperative.

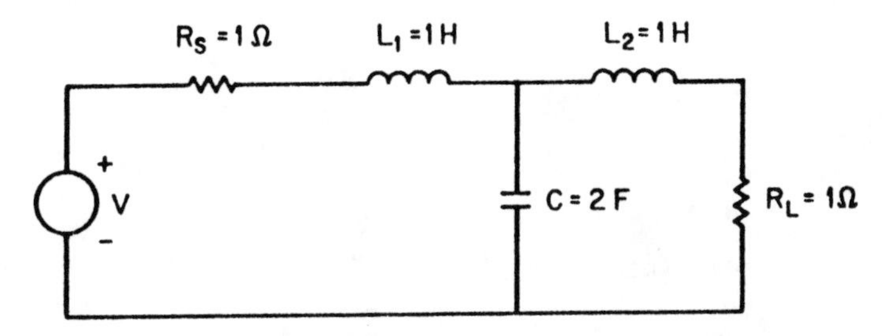

Figure 10.17 Circuit for Example 10.8.

Chapter

11

Symbolic Network Functions

All circuit analysis techniques presented up to this point apply to situations in which numerical values are available for all parameters needed to characterize the circuit elements. This chapter introduces several techniques for generation of network functions for circuits in which some of the circuit element parameters are retained as variables in symbolic form. These techniques prove very useful for large-change sensitivity analysis and for some iterative design strategies.

11.1 Signal Flow Graphs

Directed graphs, discussed in Sec. 8.1, can be used to graphically represent a system of linear equations. Specifically, we examine a modified form of directed graph called a *signal flow graph* (SFG). The modification consists of associating weights with each node and each edge. The node weights correspond to edge variables, and the edge weights correspond to equation coefficients. To determine the system of equations represented by a given SFG, construct an equation for each node by setting the node variable equal to the sum of the incoming edge terms, where each edge term is the product of the edge coefficient and the node variable for the edge's node of origin. A node that has only outgoing edges is called a *source node*, and a node with one or more incoming edges is called a *dependent node*. The definitions of *loop* and *path* within an SFG are identical to the corresponding definitions for directed graphs. (See Sec. 8.1.) The loop weight of a closed loop in an SFG is the product of all the individual edge weights contained in the loop. Likewise, the path weight of a path in an SFG is the product of all the individual edge weights contained in the path.

Example 11.1 Determine the equations depicted by the signal flow graph shown in Fig. 11.1. Identify source nodes and dependent nodes.

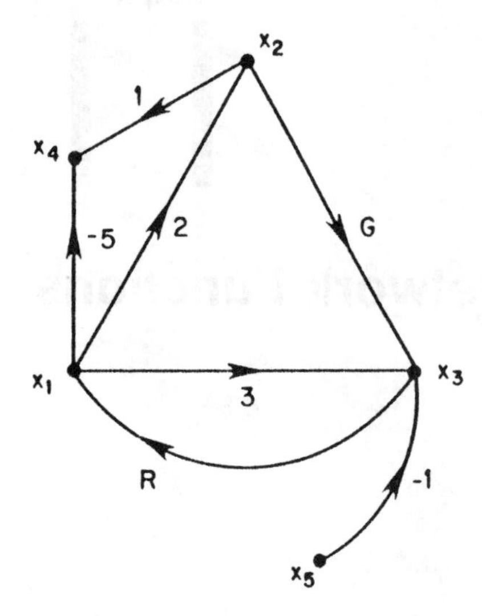

Figure 11.1 Signal flow graph for Example 11.1.

solution Node x_5 is a source node. Nodes x_1, x_2, x_3, and x_4 are dependent nodes. The equations depicted are as follows:

$$x_1 = Rx_3 \qquad x_2 = 2x_1$$

$$x_3 = 3x_1 + Gx_2 - x_5 \qquad x_4 = x_2 - 5x_1$$

The following algorithm provides a systematic approach for generating signal flow graphs for networks containing resistors, capacitors, inductors, conductors, independent sources, and controlled sources.

Algorithm 11.1 (Generating signal flow graphs for *RLC* circuits with independent and dependent sources)

Step 1. Replace each VCVS or CCVS with an independent voltage source. Use a dummy variable to represent the voltage of the IVS.

Step 2. Replace each VCCS or CCCS with an independent current source. Use a dummy variable to represent the current of the ICS.

Step 3. Form a graph corresponding to the modified circuit.

Step 4. From this graph, select a tree such that all voltage sources are contained in twigs (tree branches) and all current sources are contained in links (cotree branches).

Step 5. For each link (cotree branch) k, the branch current I_k can be expressed as $I_k = Y_k V_k$, where Y_k is the link admittance. By using the KVL, the link voltage V_k can be expressed in terms of twig (tree branch) voltages. Combine these two relationships to express the link current I_k in terms of the link admittance Y_k and appropriate twig voltages:

$$I_k = f_k(Y_k, \mathbf{V}_T)$$

Step 6. For each twig (tree branch) j, the branch voltage V_j can be expressed as $V_j = Z_j I_j$, where Z_j is the twig impedance. By using the KCL, the twig current I_j can be expressed in terms of link (cotree branch) currents. Combine these two relationships to express the twig voltage V_j in terms of the twig impedance Z_j and appropriate link currents:

$$V_j = f_j(Z_j, \mathbf{I}_L)$$

Step 7. Put into the SFG one node for each twig voltage and one node for each link current. Connect these nodes with appropriately weighted edges as needed to depict the equations for I_k and V_j found in steps 5 and 6.

Step 8. Express desired outputs in terms of twig voltages and link currents. Add the corresponding nodes and edges to the SFG.

Step 9. Express controlling variables (for controlled sources) in terms of twig voltages and link currents. Add the corresponding nodes and edges to the SFG.

Step 10. Add edges to the SFG to depict the original relationship between the controlled variable and the dummy variable introduced in steps 1 and 2. ■

Example 11.2 Use Algorithm 11.1 to generate the signal flow graph for the circuit shown in Fig. 11.2.

solution

Step 1 The VCVS connected between nodes 0, 3, and 4 is replaced with an independent voltage source connected between nodes 0 and 4, as shown in Fig. 11.3. The voltage of this source is denoted by V_x.

Step 2 There are no controlled current sources in the given circuit.

Step 3 The graph corresponding to the modified circuit is shown in Fig. 11.4.

Step 4 Edges a and f contain voltage sources and must therefore be included in the tree. We select edges ($abcf$) for the tree and edges (de) for the cotree, as shown in Fig. 11.5.

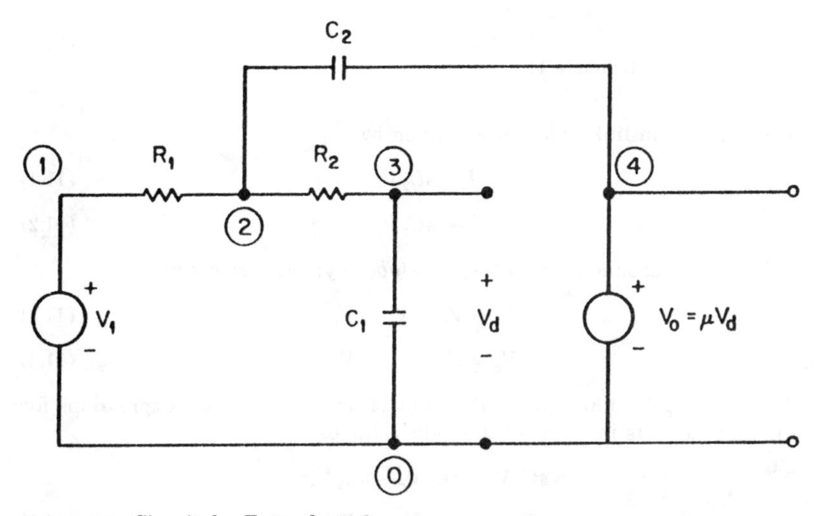

Figure 11.2 Circuit for Example 11.2.

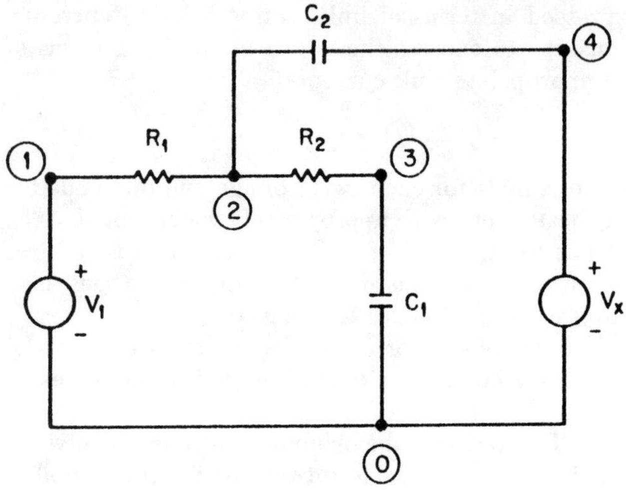

Figure 11.3 Circuit for Example 11.2 after removal of controlled sources.

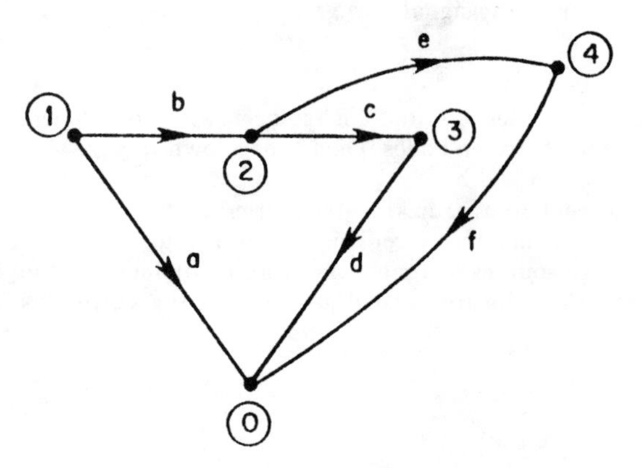

Figure 11.4 Graph for circuit of Example 11.2.

Step 5 The currents in links *e* and *d* are given by

$$I_e = sC_2 V_e \tag{11.1}$$

$$I_d = sC_1 V_d \tag{11.2}$$

KVL applied around loops (*abef*) and (*abcd*) yields, respectively,

$$V_e = V_a - V_b - V_x \tag{11.3}$$

$$V_d = V_a - V_b - V_c \tag{11.4}$$

Substitution of (11.3) into (11.1) and (11.4) into (11.2) yields expressions for the link currents in terms of the twig voltages:

$$I_e = sC_2 V_a - sC_2 V_b - sC_2 V_x \tag{11.5}$$

$$I_d = sC_1 V_a - sC_1 V_b - sC_1 V_c \tag{11.6}$$

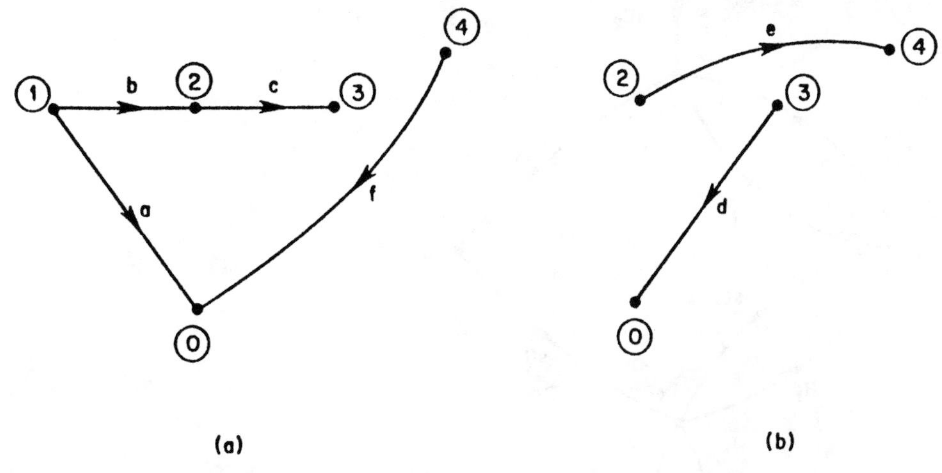

Figure 11.5 Graph for Example 11.2 after separation into (*a*) tree and (*b*) cotree subgraphs.

Step 6 The voltages in twigs b and c are given by

$$V_b = R_1 I_b \tag{11.7}$$

$$V_c = R_2 I_d \tag{11.8}$$

KCL applied at nodes 2 and 3 yields

$$I_b = I_c + I_e \tag{11.9}$$

$$I_c = I_d \tag{11.10}$$

Equation (11.9) is not suitable "as is" because I_c is a twig current rather than a link current. However, we can substitute (11.10) into (11.9) to obtain

$$I_b = I_d + I_e \tag{11.11}$$

Finally, substitution of (11.11) into (11.7) and (11.10) into (11.8) yields expressions for the link currents in terms of the twig voltages:

$$V_b = R_1 I_d + R_1 I_e \tag{11.12}$$

$$V_c = R_2 I_d \tag{11.13}$$

Step 7 The partial SFG depicting Eqs. (11.5), (11.6), (11.12), and (11.13) is shown in Fig. 11.6.

Step 8 Assuming that V_0 is the desired output, a node does not need to be added to the SFG because $V_0 = V_x$ and a node for V_x already exists.

Step 9 The controlling variable V_d is a link voltage and therefore has not yet been included in the SFG. From KVL applied around loop (*abcd*) we determine that $V_d = V_a - V_c - V_b$ and add the appropriate edges, as shown in Fig. 11.7.

Step 10 The controlled voltage V_x is equal to μV_d. This relationship is incorporated into the SFG by adding an edge of weight μ from node V_d to node V_x. The completed SFG is shown in Fig. 11.8.

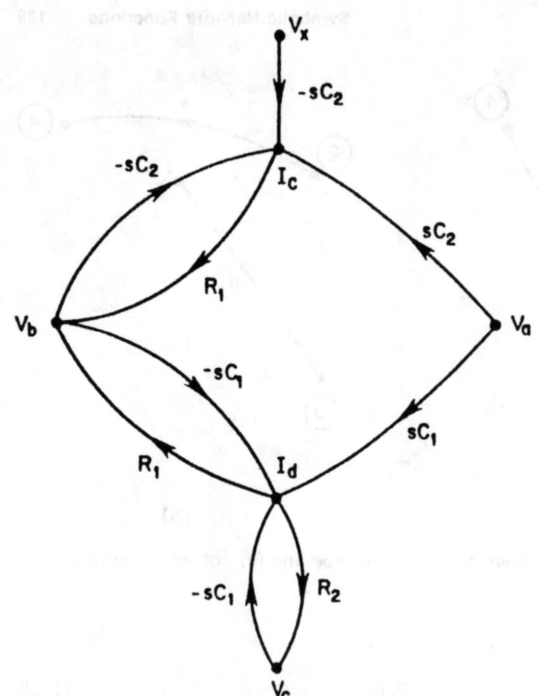

Figure 11.6 Partial SFG depicting equations for twig voltages and link currents.

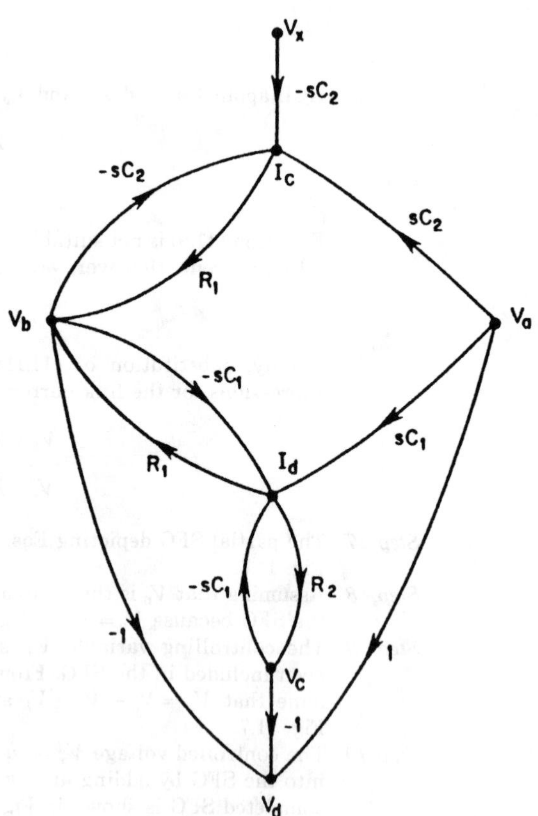

Figure 11.7 Partial SFG after addition of controlling variables.

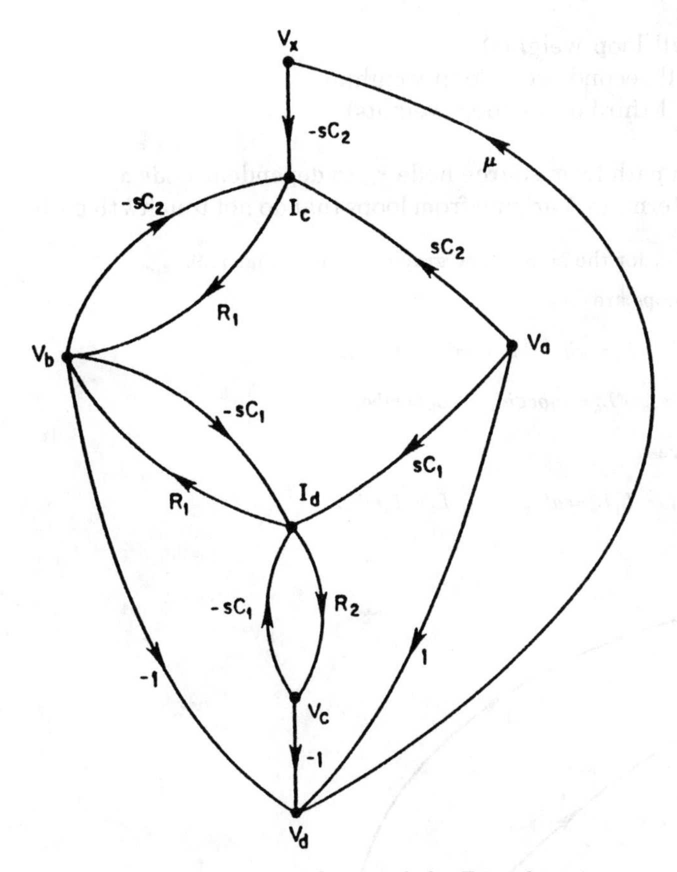

Figure 11.8 Completed signal flow graph for Example 11.2.

11.2 Mason's Rule

Let x_{si} and x_j denote, respectively, the ith source node variable and jth dependent-node variable in the signal flow graph for a linear network. For a linear network having m source nodes, any dependent-node variable x_j can be expressed as a linear combination of the source node variables:

$$x_j = T_{j1}x_1 + T_{j2}x_{s2} + \cdots + T_{jn}x_{sm}$$

where the coefficient T_{ji} is known as the *transmission* from source node x_{si} to dependent node x_j. *Mason's rule* provides a means for computing the various T_{ji} values:

$$T_{ji} = \frac{1}{\Delta}\sum_k P_k\Delta_k$$

where $\Delta = 1 -$ (sum of all loop weights)
$+$(sum of all second-order loop weights)
$-$(sum of all third-order loop weights)
...

P_k = weight of kth path from source node x_{si} to dependent node x_j
Δ_k = sum of those terms in Δ arising from loops that do not touch kth path

Example 11.3 Compute T_{XA} for the signal flow graph shown in Fig. 11.9.

solution The first-order loops are

$$L_1 = ab \qquad L_2 = cd \qquad L_3 = fg$$

$$L_4 = mbcghj \qquad L_5 = mbej$$

The second-order loops are

$$L_1 L_3 = abfg \qquad L_3 L_5 = fgmbej$$

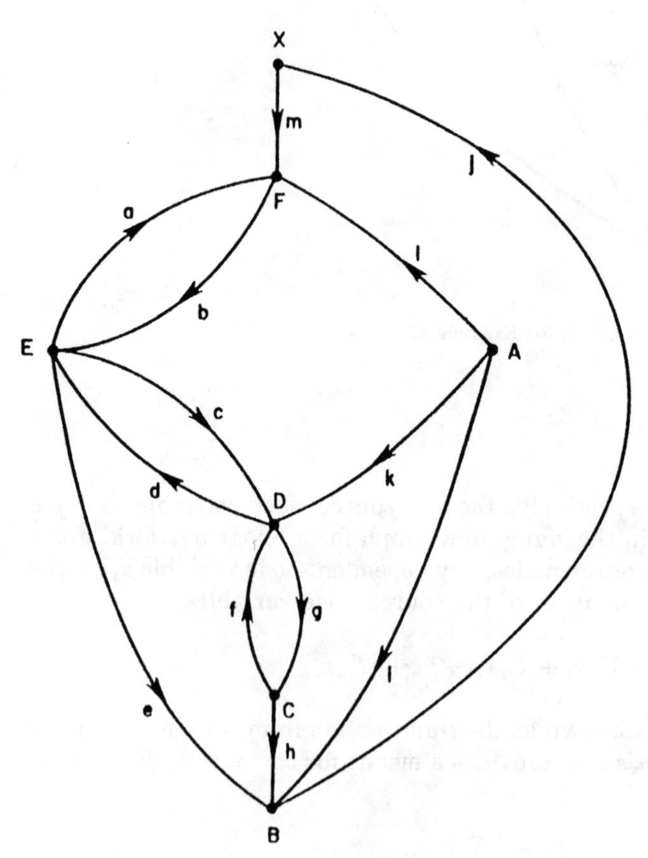

Figure 11.9 Signal flow graph for Example 11.3.

There are no third- or higher-order loops. Therefore,

$$\Delta = 1 - ab - cd - fg - mbcghj - mbej + abfg + fgmbej$$

The five paths from node A to node X are

$$P_1 = kghj \qquad P_2 = kdej \qquad P_3 = ij \qquad P_4 = lbcghj \qquad P_5 = lbej$$

Careful examination of Fig. 11.9 reveals that

P_2 and P_4 each touch all loops.
P_1 touches all loops except L_1.
P_3 touches all loops except L_1, L_2, L_3, and L_1L_3.
P_5 touches all loops except L_3.

Therefore, $\Delta_1 = 1 - ab \qquad \Delta_2 = \Delta_4 = 1 \qquad \Delta_5 = 1 - fg$

$$\Delta_3 = 1 - ab - cd - fg + abfg$$

$$T_{XA} = \frac{(kghj)(1 - ab) + kdej + lbcghj + ij(1 - ab - cd - fg + abfg) + lbej(1 - fg)}{1 - ab - cd - fg - mbcghj - mbej + abfg + fgmbej}$$

11.3 Tree Enumeration Method

Assume that we have a network of n nodes that has a nodal admittance matrix $\mathbf{Y}$. Further assume that we treat this network as a two-port by connecting an input signal source between node 1 and the reference node. The output voltage is measured between node 2 and the reference node. (The node numbering should be arranged so that the input and output nodes are nodes 1 and 2, respectively.) The input impedance, transfer impedance, and transfer voltage gain can be computed as

$$Z_{\text{in}} = \frac{V_{\text{in}}}{I_{\text{in}}} = \frac{V_1}{I_1} = \frac{\Delta_{11}}{\Delta} \tag{11.14}$$

$$\frac{V_0}{I_{\text{in}}} = \frac{V_2}{I_1} = \frac{\Delta_{12}}{\Delta} \tag{11.15}$$

$$\frac{V_0}{V_{\text{in}}} = \frac{V_2}{V_1} = \frac{\Delta_{12}}{\Delta_{11}} \tag{11.16}$$

where Δ denotes the determinant of $\mathbf{Y}$ and Δ_{ij} denotes the cofactor, defined as

$$\Delta_{ij} \triangleq (-1)^{i+j}\det[\mathbf{M}_{ij}]$$

The matrix $\mathbf{M}_{ij}$ is formed by removing row i and column j from $\mathbf{Y}$. Equations (11.14), (11.15), and (11.16) can obviously be applied to an all-numeric matrix $\mathbf{Y}$, but with a bit of finagling they can also be applied to nodal admittance matrices that contain symbolic entries.

Algorithm 11.2 (Tree enumeration method for generation of symbolic network functions)

Step 1. Augment the network by adding an admittance y and a voltage-controlled current source $g_s V_0$ across the input port.

Step 2. Form an indefinite nodal admittance matrix $\mathbf{Y}_{\text{ind}}$ from the usual nodal admittance matrix $\mathbf{Y}$ by adding a row and column for the reference node. The appropriate entries for the new row and new column are easily found by exploiting the fact that the sum of the elements in each row or each column is zero.

Step 3. Form the directed graph corresponding to $\mathbf{Y}_{\text{ind}}$. For the entry in row i, column j, the corresponding edge is directed from node i to node j in the graph and has a weight equal to -1 times the matrix entry. Diagonal elements of the matrix do not explicitly appear in the graph.

Step 4. Select any node in the graph, and list (enumerate) all the directed trees having the selected node as a root.

Step 5. For each tree, compute the product of all edge weights in the tree. Compute D as the sum of these products for all trees found in step 3.

Step 6. Separate the terms of D involving y_s and g_s so that

$$D = A + y_s B + g_s C$$

where A, B, and C do not contain y_s or g_s. It can be shown that

$$\Delta = A \qquad \Delta_{11} = B \qquad \Delta_{12} = C$$

Step 7. Evaluate (11.14), (11.15), or (11.16) as desired, using the values for Δ, Δ_{11}, and Δ_{12} obtained in step 6. ■

Example 11.4 Use Algorithm 11.2 to compute Z_{in}, V_0/V_{in}, and V_0/I_{in} for the circuit shown in Fig. 11.10.

solution The indefinite nodal admittance matrix is given by

$$\mathbf{Y}_{\text{ind}} = \begin{bmatrix} \dfrac{1}{R_1} + y_s & g_s & \dfrac{-1}{R_1} & -y_s - g_s \\ 0 & \dfrac{1}{R_2} & \dfrac{-1}{R_2} & 0 \\ \dfrac{-1}{R_1} & \dfrac{-1}{R_2} & \dfrac{1}{R_1} + \dfrac{1}{R_2} + sC & -sC \\ -y_s & -g_s & -sC & sC + y_s + g_s \end{bmatrix}$$

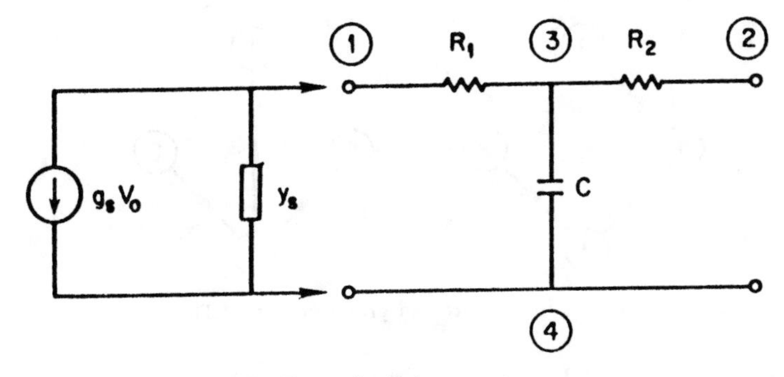

Figure 11.10 Circuit for Example 11.4.

The corresponding directed graph is shown in Fig. 11.11. Node 2 only has one outgoing edge, so using it as the root will likely result in the fewest number of trees to be evaluated. Figure 11.12 shows the 3 trees rooted at node 2, and for comparison Fig. 11.13 shows the 12 trees rooted at node 4. The determinant D is given by

$$D = \frac{1}{R_1}\frac{1}{R_2}(y_s + g_s) + \frac{1}{R_2}sCy_s + \frac{1}{R_2}sC\frac{1}{R_1}$$

$$= \frac{sC}{R_1R_2} + \frac{y_s(R_1sC+1)}{R_1R_2} + \frac{g_s}{R_1R_2}$$

$$\Delta = \frac{sC}{R_1R_2} \qquad \Delta_{11} = \frac{R_1sC+1}{R_1R_2} \qquad \Delta_{12} = \frac{1}{R_1R_2}$$

$$Z_{\text{in}} = \frac{R_1sC+1}{sC} \qquad \frac{V_0}{V_{\text{in}}} = \frac{1}{R_1sC+1} \qquad \frac{V_0}{I_{\text{in}}} = \frac{1}{sC}$$

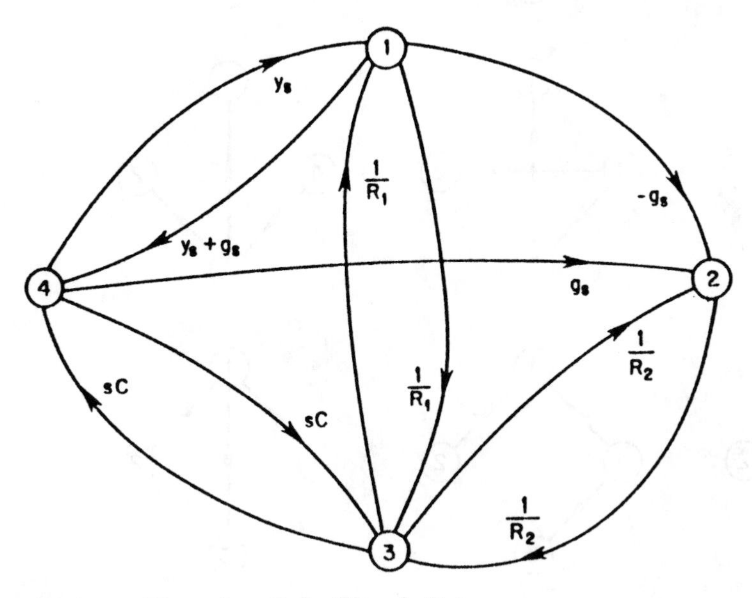

Figure 11.11 Directed graph for Example 11.4.

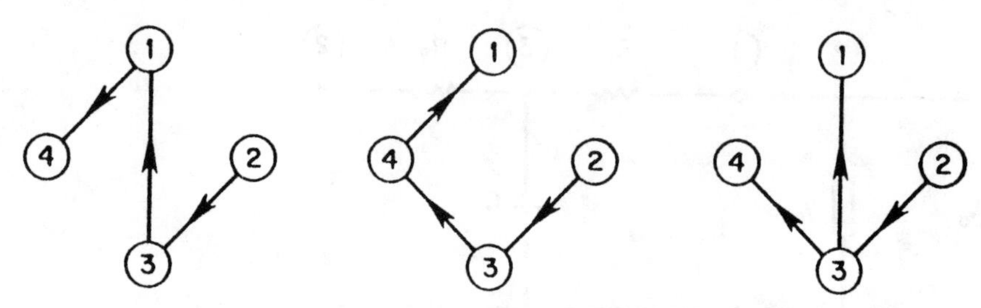

Figure 11.12 The three trees rooted at node 2 in the directed graph of Fig. 11.11.

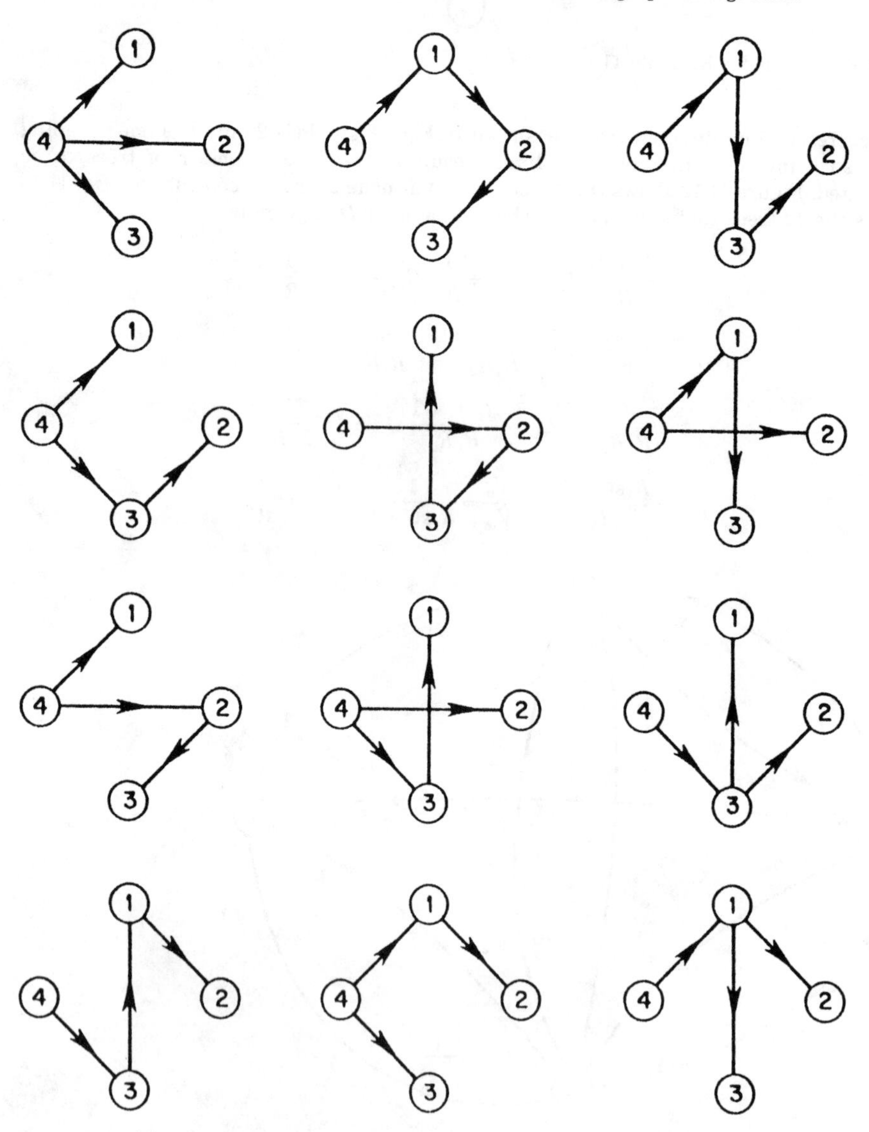

Figure 11.13 The 12 trees rooted at node 4 in the directed graph of Fig. 11.11.

11.4 Parameter Extraction Method

The tree enumeration method of Sec. 11.3 is fine for small circuits, but the number of trees quickly gets out of hand as the circuit size increases. The *parameter extraction* method is a more manageable, matrix-based approach for evaluating the determinants and cofactors needed for Eqs. (11.14), (11.15), and (11.16). A useful property of the indefinite nodal admittance matrix $\mathbf{Y}_{\text{ind}}$ is that all cofactors of $\mathbf{Y}_{\text{ind}}$ are equal. Therefore, we can speak of the "cofactor of $\mathbf{Y}_{\text{ind}}$" without specifying a row and column to be deleted, because it doesn't matter which row and column are deleted.

Algorithm 11.3 (Computing the cofactor for an indefinite nodal admittance matrix having symbolic entries). This algorithm assumes that any symbolic term α appearing in the indefinite nodal admittance matrix appears in a total of four locations: α appears in row i, column k and row j, column m; $-\alpha$ appears in row i, column m and row j, column k.

Step 1. Evaluate any cofactor of $\mathbf{Y}_{\text{ind}}$ for $\alpha = 0$. Call the result a.

Step 2. Add row j to row i.

Step 3. Add column m to column k.

Step 4. Delete row j and column m.

Step 5. Evaluate any cofactor of $\mathbf{Y}_{\text{ind}}$ as modified by steps 2, 3, and 4. Call the result b.

Step 6. Obtain the cofactor of the original $\mathbf{Y}_{\text{ind}}$, using

$$\text{cof}(\mathbf{Y}_{\text{ind}}) = a + (-1)^{j+m}\alpha b$$ ■

Steps 1 and 5 of the algorithm call for evaluation of a cofactor. If all the symbolic entries have been extracted, then this evaluation is straightforward. However, if symbolic entries remain, the algorithm must be applied recursively until an all-numeric matrix remains.

Chapter

12

Sensitivity

In general, component values differ—often significantly—from their nominal values. A circuit constructed from actual components may exhibit performance that differs greatly from that predicted by an analysis based on nominal component values. Sensitivity analysis is a valuable tool for determining just how much a circuit's performance will vary as a function of deviation in component values. This knowledge allows a designer to determine which portions of a circuit need to use high-precision components and which portions can use lower-precision (and hence less expensive) components.

12.1 Sensitivity Concepts

Intuitively, the sensitivity of a network function T with respect to a parameter x is given by the partial derivative of T with respect to x, or $\partial T/\partial x$. It is customary practice, however, to define the sensitivity of T with respect to x as

$$S_x^T \triangleq \frac{\partial T/T}{\partial x/x} = \frac{\partial T}{\partial x} \cdot \frac{x}{T} = \frac{\partial \ln T}{\partial \ln x} \tag{12.1}$$

To emphasize the distinction between S_x^T and $\partial T/\partial x$, the former is called the *relative sensitivity* or *normalized sensitivity* while the latter is sometimes called the *unnormalized sensitivity.* Often we will be interested in the sensitivity of a network function with respect to a parasitic parameter that is nominally zero. In this case, to avoid division by zero, the sensitivity must be defined as

$$\mathcal{S}_x^T \triangleq \frac{\partial T/T}{\partial x} = \frac{\partial T}{\partial x} \cdot \frac{1}{T} = \frac{\partial \ln T}{\partial x} \tag{12.2}$$

12.2 Incremental-Network Approach

The *incremental-network* approach for sensitivity analysis is based upon the idea of constructing a new network that has the same topology as the original network, but in which each current I_n of the original is replaced by an incremental current ΔI_n and each voltage V_n of the original is replaced by an incremental voltage ΔV_n. This new network can then be solved to obtain the incremental (as opposed to differential) sensitivity of any parameter with respect to any other.

Exactly how do we construct the incremental network? Let's consider the specific case of an impedance Z. As always, the current and voltage are related via Ohm's law:

$$V = ZI$$

If the impedance is changed by an amount ΔZ, the voltage and current will also change:

$$\begin{aligned} V + \Delta V &= (Z + \Delta Z)(I + \Delta I) \\ &= ZI + Z\,\Delta I + \Delta Z\,I + \Delta Z\,\Delta I \end{aligned} \tag{12.3}$$

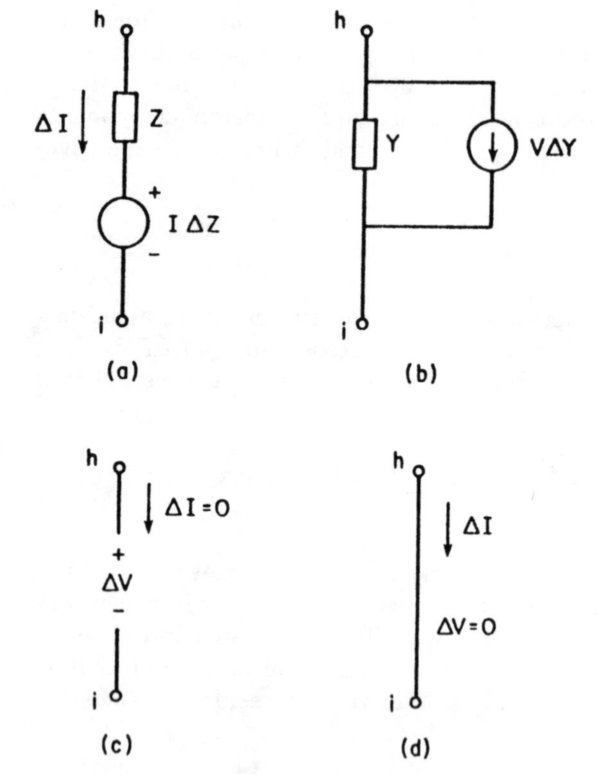

Figure 12.1 Incremental network configurations for (*a*) impedance, (*b*) admittance, (*c*) voltage source, (*d*) current source.

Since $V = ZI$, we can subtract V from the LHS and ZI from the RHS of Eq. (12.3) to obtain

$$\Delta V = Z\,\Delta I + I\,\Delta Z + \Delta Z\,\Delta I$$

At this point, we commence handwaving and make the assumption that since ΔI and ΔZ are tiny, the product $\Delta Z\,\Delta I$ is "tiny squared" and can therefore be neglected, so that

$$\Delta V \cong Z\,\Delta I + I\,\Delta Z \tag{12.4}$$

From (12.4) we conclude that a circuit branch as shown in Fig. 12.1*a*, containing an impedance of Z in series with a voltage source of $I\,\Delta Z$, will admit a current of ΔI and exhibit a voltage drop of ΔV. Thus every impedance in the original network will be replaced in the incremental network by an impedance in series with a voltage source. Similar developments can be used to obtain the corresponding incremental-network branches for admittances, dependent sources, and independent sources.

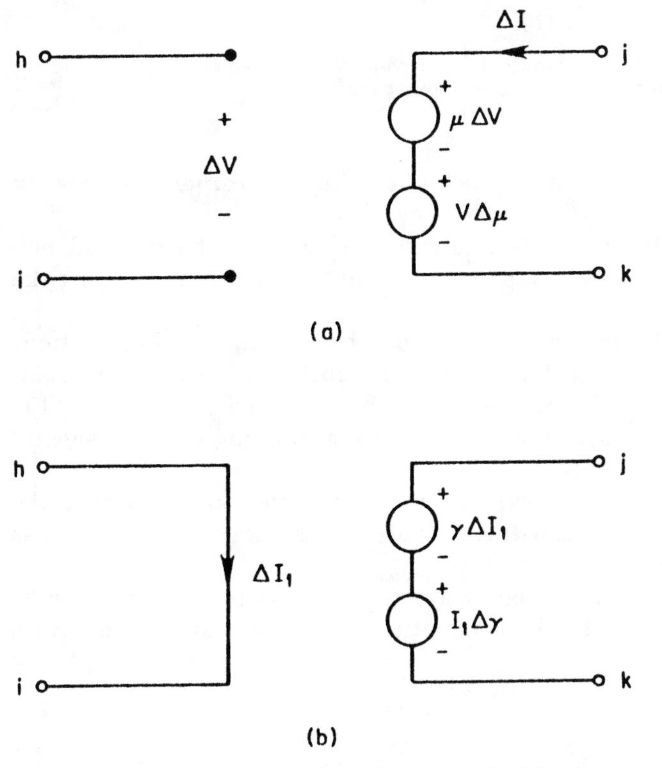

Figure 12.2 Incremental network configurations for (*a*) voltage-controlled voltage source (VCVS) and (*b*) current-controlled voltage source (CCVS).

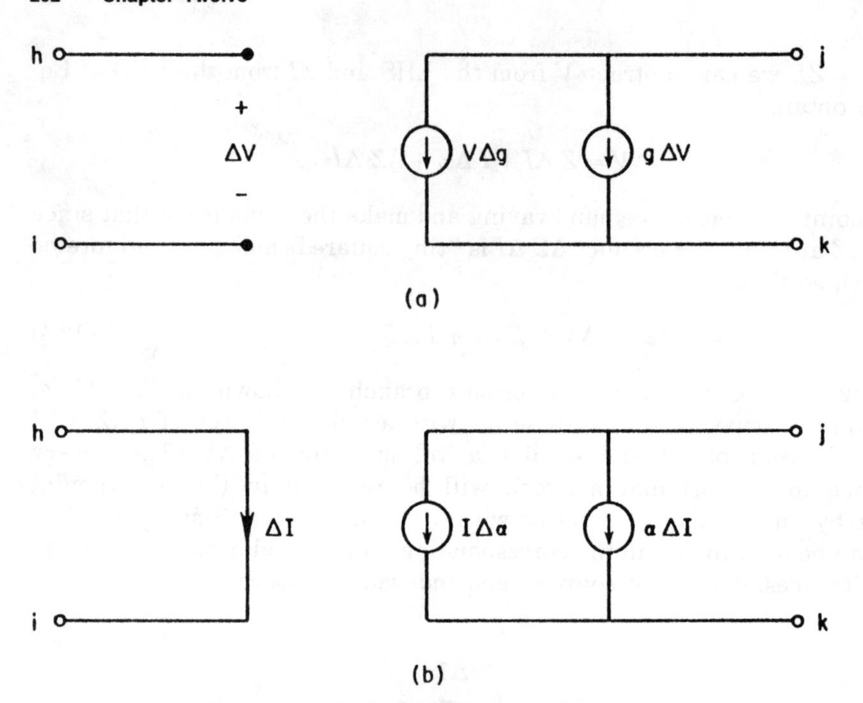

Figure 12.3 Incremental network configurations for (*a*) voltage-controlled current source (VCCS) and (*b*) current-controlled current source (CCCS).

Algorithm 12.1 (Sensitivity analysis via the incremental-network method)

Step 1. Formulate the network (matrix) equation for the original network, using the techniques of Chap. 9, and solve this equation by using the techniques of Chap. 7.

Step 2. Construct the incremental network by making the substitutions indicated in Figs. 12.1 through 12.3. (*Note:* The labeling of nodes and circuit parameters in these figures is consistent with Figs. 9.5 and 9.7 through 9.10.)

Step 3. Formulate the symbolic network equations for the incremental network.

Step 4. Substitute the known values for all terms that also appear in the original network equation. This will leave only incremental variables such as ΔV, ΔI, ΔZ, and Δg.

Step 5. The sensitivity of any parameter with respect to any other can be approximated by solving the symbolic equations to form the appropriate ratio of incremental parameters. ■

Chapter

13

Design Studies

This chapter shows how the various techniques presented in earlier chapters can be pulled together and used in an "end-to-end" design effort. A single common design study is used to illustrate each of the different methods. Assume that we must design a lowpass filter with no more than 0.5-dB variation in the magnitude response across a passband of 3 kHz. Attenuations of 40 and 60 dB are required at 6 and 9 kHz, respectively. Furthermore, the filter's load and source terminations must each be approximately 1000 Ω.

13.1 Approximation and Synthesis

Examination of Fig. 6.30 reveals that a fifth-order Chebyshev filter with 0.5-dB ripple should be able to satisfy our stated requirements. We will use Algorithm 10.7 to synthesize a double-terminated network of the form shown in Fig. 13.1. Step 1 of the algorithm requires that we obtain

$$F(s) = \left|H(j\omega)\right|^2\Big|_{\omega = s/j}$$

which from Eq. (6.7) we know to be given by

$$F(s) = \frac{1}{1 + \epsilon^2 T_5^2(\omega)}\Bigg|_{\omega = s/j}$$

where $\epsilon^2 = 10^{0.5/10} - 1 = 0.12202$. From Table 6.2, we obtain $T_5(\omega)$ as

$$T_5(\omega) = 16\omega^2 - 20\omega^3 + 5\omega$$

and $T_5^2(s)$ is subsequently obtained as

$$T_5(s) = T_5(\omega)\big|_{\omega = s/j} = \frac{1}{j}\,(16s^5 + 20s^3 + 5s)$$

$$T_5^2(s) = -(16s^5 + 20s^3 + 5s)^2$$

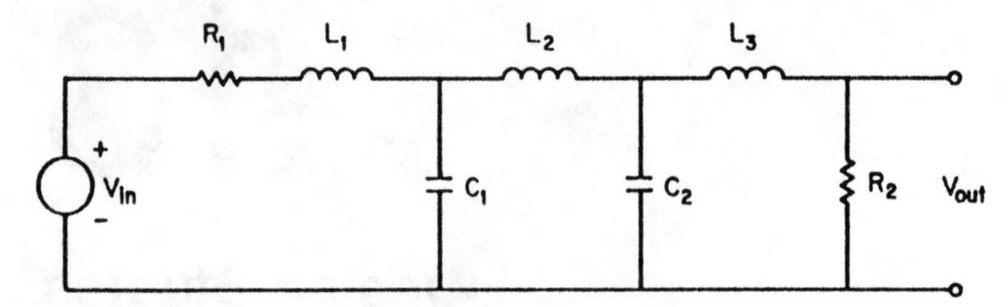

Figure 13.1 Circuit configuration for a fifth-order filter.

Let $R_1 = R_2 = 1$. Then for step 3 of Algorithm 10.7 we obtain

$$\frac{A(s)}{B(s)} = 1 - \frac{4R_1R_2}{(R_1+R_2)^2}F(s)$$

$$= \frac{\epsilon^2 T_5^2(s)}{1+\epsilon^2 T_5^2(s)} = \frac{T_5^2(s)}{T_5^2(s)+1/\epsilon^2}$$

$$= \frac{(16s^5+20s^3+5s)^2}{(16s^5+20s^3+5s)^2 - 8.1955}$$

The roots of $A(s) = (16s^5 + 20s^3 + 5s)^2 = 0$ are 0, $\pm 0.951057j$, and $\pm 0.587785j$, thus yielding

$$p(s) = s^5 + 1.25s^3 + 0.3125s$$

The roots of $B(s)$ are found to be

$$s_1 = -0.111963 + 1.011557j \qquad s_6 = 0.111963 - 1.011557j$$

$$s_2 = -0.293123 + 0.625177j \qquad s_7 = 0.293123 - 0.625177j$$

$$s_3 = -0.36232 \qquad s_8 = 0.36232$$

$$s_4 = -0.293123 - 0.625177j \qquad s_9 = 0.293123 + 0.625177j$$

$$s_5 = -0.111963 - 1.011557j \qquad s_{10} = 0.111963 + 1.011557j$$

Roots s_1, s_2, s_3, s_4, and s_5 lie in the left half-plane and are therefore assigned to $q(s)$:

$$q(s) = (s-s_1)(s-s_2)(s-s_3)(s-s_4)(s-s_5)$$

$$= s^5 + 1.172492s^4 + 1.937368s^3 + 1.309576s^2 + 0.752518s + 0.178923$$

We next perform step 8 of the synthesis algorithm by computing $Z(s)$ as

$$Z(s) = \frac{q(s)+p(s)}{q(s)-p(s)}$$

$$= \frac{2s^5 + 1.172492s^4 + 3.187368s^3 + 1.309576s^2 + 1.065018s + 0.178923}{1.172492s^4 + 0.687368s^3 + 1.309576s^2 + 0.440018s + 0.178923}$$

Continued-fraction expansion of $Z(s)$ yields

$$Z(s) = 1.706s + \cfrac{1}{1.23s + \cfrac{1}{2.541s + \cfrac{1}{1.23s + \cfrac{1}{1.706s + 1}}}}$$

so we have

$$L_1 = L_3 = 1.706 \text{ H} \qquad L_2 = 2.541 \text{ H} \qquad C_1 = C_2 = 1.23 \text{ F} \tag{13.1}$$

To scale the filter for a ripple bandwidth of 3 kHz, each inductor and capacitor value must be divided by 3000. Impedance scaling to make $R_1 = R_2 = 1000$ is accomplished by dividing each capacitor value by 100 and multiplying each inductor value by 1000. After both frequency scaling and impedance scaling have been performed, the resulting component values are

$$\begin{gathered} L_1 = L_3 = 5.6854 \times 10^{-1} \text{ H} \qquad L_2 = 8.4694 \times 10^{-1} \text{ H} \\ C_1 = C_2 = 4.0987 \times 10^{-7} \text{ F} \end{gathered} \tag{13.2}$$

Alternatively, the filter can be realized by the circuit of Fig. 13.2 with component values of

$$\begin{gathered} C_1 = C_3 = 5.6854 \times 10^{-7} \text{ H} \qquad C_2 = 8.4694 \times 10^{-7} \text{ H} \\ L_1 = L_2 = 4.0987 \times 10^{-1} \text{ F} \end{gathered} \tag{13.3}$$

13.2 Performance

We can check the synthesis results of the previous section by using the C program given in Listing 13.1. The responses of both filters are very close to ideal, but we would have some trouble actually building either circuit. Nearly ideal resistors and capacitors could be obtained, but the exact values indicated by the synthesis are not readily available.

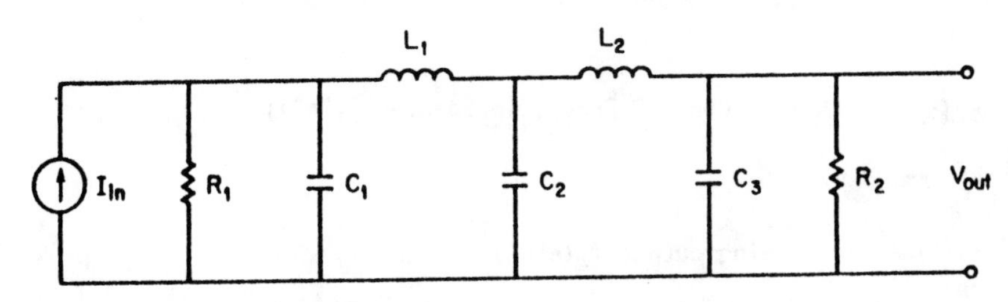

Figure 13.2 Alternative circuit configuration for a fifth-order filter.

```
/***************************************/
/*                                     */
/*   Listing 13.1                      */
/*                                     */
/*   Main Program for Evaluating       */
/*   Frequency Response of a Circuit   */
/*                                     */
/***************************************/
#include "globDefs.h"
#include "protos.h"

FILE *inFile;
FILE *outFile;

main()
{
real freq;
static struct complex Y[20][20];
static struct complex Y_of_s[20][20];
static struct complex J_vector[20], J_work[20];
static struct complex xVector[20];
static struct complex LU_matrix[20][20];
static char dependentVariable[20][5];
static int d[20];
real radFreq, normFreq;
int order, numBranches, i, index_org, iFreq, pointsPerDecade;
int returnCode;

printf("normalization frequency ?\n");
scanf("%lg",&normFreq);
printf("points per decade ?\n");
scanf("%d",&pointsPerDecade);
inFile = fopen("CircSpec.inp","r");
if( inFile == NULL)
    {
    printf("failure opening input file\n");
    return;
    }
outFile = fopen("C System:Chap 13 Progs:CircSpec.out","w");

if( outFile == NULL)
    {
    printf("failure opening output file\n");
    return;
    }
```

```
fprintf(outFile,"normFreq = %f\n",normFreq);
fprintf(outFile,"pointsPerDecade = %d\n",pointsPerDecade);

BuildAugNodalEqn( Y_of_s, J_vector, &order, dependentVariable);

for(iFreq=1; iFreq<=2 * pointsPerDecade; iFreq++)
    {
    radFreq = pow(TEN, (double) ((iFreq-pointsPerDecade)/((double)
                pointsPerDecade)));
    freq = normFreq * radFreq/TWO_PI;
    SetFreq( freq, Y_of_s, order, Y);
    for(i=1; i<=order; i++) J_work[i] = J_vector[i];

    DoolittleMethod( order, Y, d, J_work);

    SubstituteLU( order, J_work, Y, xVector,d);

    fprintf(outFile,"%d) freq = %f mag x3 = %e\n", iFreq,
            TWO_PI * freq, 20.0 * log10(2.0*cAbs(xVector[3])));
    }

fclose(outFile);
return;
}
```

Standard values

Let's modify the circuit of Fig. 13.2 to use standard 5 percent resistor and capacitor values by setting

$$C_1 = C_3 = 5.6 \times 10^{-7} \qquad C_2 = 8.2 \times 10^{-7}$$

The magnitude responses of this circuit and the "ideal-value" circuit are compared in Table 13.1. From the values tabulated, we note that the passband ripple exceeds specifications by 0.0357 dB and that the attenuations at 6 and 9 kHz each fall short by about 0.6 dB.

So far, we have not considered the difference between the components' nominal and actual values. Each component could weigh in at full tolerance low, full tolerance high, or at any value between these two extremes. If R_1 of the circuit in Fig. 13.2 is 5 percent below its nominal value—that is, 950 versus 1000 Ω—the passband response will be as shown in Fig. 13.3. The attenuations at 6 and 9 kHz are not too bad, but the passband ripple reaches 0.75 dB. On the other hand, if the two resistor values are 5 percent high while the capacitors are each 5 percent low, the passband ripple will be limited to approximately 0.42 dB but the attenuations at 6 and 9 kHz will each be about

TABLE 13.1 Comparison of Filter Performance with Ideal Component Values and with Standard Values

Frequency (kHz)	Attenuation with ideal components (dB)	Attenuation with standard components (dB)
0.3	0.12049	0.11684
0.6	0.36289	0.33737
0.9	0.49894	0.46500
1.2	0.39540	0.36851
1.5	0.13050	0.12039
1.8	0.00305	0.00370
2.1	0.23219	0.23751
2.4	0.49728	0.53570
2.7	0.20677	0.29934
3.0	0.50000	0.23670
6.0	42.0387	41.4219
9.0	61.3988	60.8263
12.0	74.4576	73.8976
15.0	84.4027	83.8482
18.0	92.4570	91.9053
21.0	99.2332	98.6832
24.0	105.085	104.536
27.0	110.236	109.688
30.0	114.838	114.290

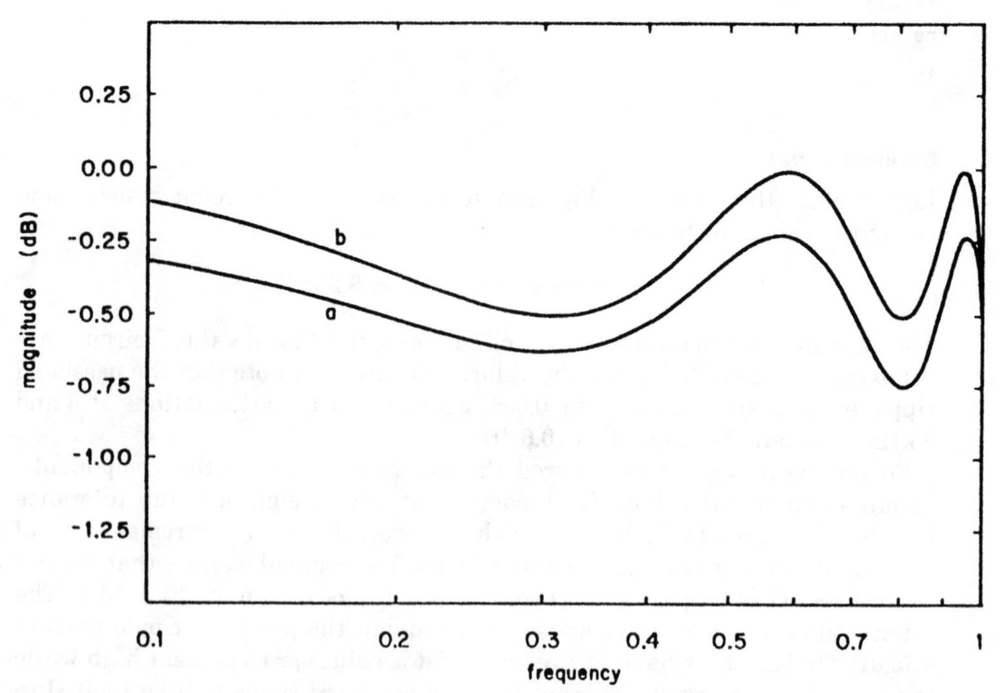

Figure 13.3 Passband magnitude response for circuit of Fig. 13.2 with the R_1 value 5 percent below nominal: (*a*) circuit response, (*b*) ideal Chebyshev response.

2 dB less than ideal. The brute-force technique for finding the worst-case filter performance involves evaluating the response for each possible combination of worst-case component values. For two resistors and three capacitors, this would involve only 32 different combinations and could be easily implemented. However, for more complicated circuits we might be better off using the techniques of Chap. 12.

13.3 Compensating for Lossy Inductors

In this book so far inductors have been modeled as ideal *lossless* inductors. As was fleetingly shown in Fig. 2.14, a practical *lossy* inductor can be modeled as an ideal inductor with a series resistor and shunt capacitor. At the frequencies of interest in this chapter, the effect of the shunt capacitor is negligible and can be ignored. However, the effects of the series resistance can be quite serious. If we model each of the inductors in the filter of Fig. 13.2 as having a series resistance of 100 Ω, we find that the magnitude varies by almost 3.5 dB across the passband, as shown in Fig. 13.4. A strategy for coping with lossy inductors involves predistorting the desired response to compensate for the effects of the series resistances. The specific procedure is given by Algorithm 13.1.

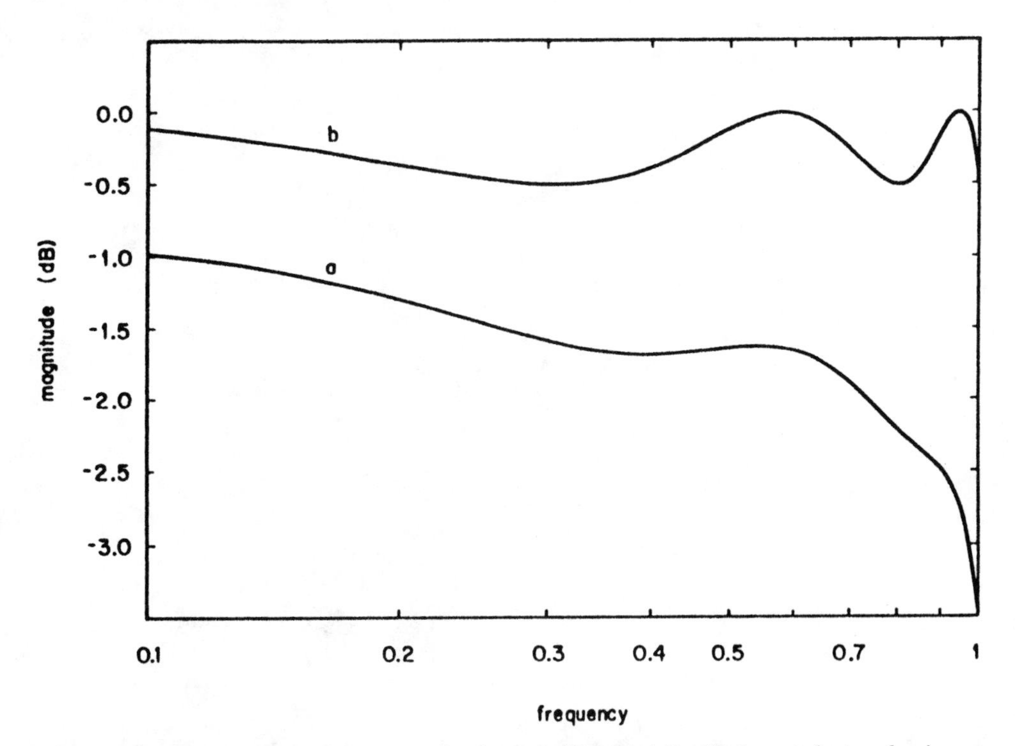

Figure 13.4 Passband magnitude response for five-pole filter circuit with lossy inductors having $P_s = 100\ \Omega$: (*a*) circuit response, (*b*) ideal Chebyshev response.

Algorithm 13.1 (Predistortion for lossy inductors)

Step 1. Synthesize the desired network function $H(s)$.

Step 2. Obtain specifications or measurements for the practical inductors to be used. For each inductor, examine the ratio of series resistance R_s to inductance L. Set d equal to the largest value of this ratio.

Step 3. Examine the poles of the desired network function $H(s)$, and determine the real part σ of the pole closest to the imaginary axis. The value of d found in step 2 must satisfy $d < |\sigma|$, in order to proceed. If this condition is not satisfied, obtain higher-Q inductors and go back to step 2.

Step 4. Calculate the network function $H(s - d)$ by substituting $s - d$ for s in $H(s)$.

Step 5. Synthesize a realization of $H(s - d)$, using ideal elements.

Step 6. In series with each inductor in the circuit realized in step 5, connect a resistance $R_x = Ld - R_s$, where R_s is the internal series resistance of the lossy inductor. (Note that for the lossiest inductor, $R_s = Ld$, so $R_x = 0$.)

Step 7. In parallel with each capacitor, connect a resistance $1/(Cd)$. ■

Appendix A

```
/***************************************/
/*                                     */
/*  Appendix A -- Global Definitions   */
/*                                     */
/*  globDefs.h                         */
/*                                     */
/*  global definitions                 */
/*                                     */
/***************************************/

#include <stdio.h>
#include <math.h>
#include <time.h>

#define EOL 10
#define STOP_CHAR 38
#define SPACE 32
#define TRUE 1
#define FALSE 0
#define PI 3.14159265
#define TWO_PI 6.2831853
#define TEN (double) 10.0
#define MAX_COLUMNS 20
#define MAX_ROWS 20

/*  structure definition for single precision complex */
/* struct complex
    {
    float Re;
```

```
     float Im;
     }; */

/*  structure definition for double precision complex */
struct complex
     {
     double Re;
     double Im;
     };

#define COMPLEX_INIT(s,a,b) {((s).Re=(a));((s).Im=(b));}


typedef int logical;
typedef double real;


enum componentTypes {
     undefined,
     resistor,
     capacitor,
     inductor,
     currentSource,
     voltageSource,
     conductance,
     twoPorts,
     VCVS,
     CCVS,
     VCCS,
     CCCS,
     transformer,
     OpAmp };

enum unitTypes {
     undefUnits,
     ohms,
     mhos,
     amps,
     farads,
     henry,
     volts};

enum responses {
     impulseResponse,
     stepResponse};
```

Appendix B

```
/**************************************************/
/*                                                */
/*  Appendix B -- Prototypes for C Functions      */
/*                                                */
/**************************************************/

int LaguerreMethod(
                        int order,
                        struct complex coef[],
                        struct complex *zz,
                        real epsilon,
                        real epsilon2,
                        int maxIterations);

int BuildNodalEqn(      real freq,
                            struct complex Y[][MAX_COLUMNS],
                            struct complex J_vector[],
                            int *num_nodes);

int BuildAugNodalEqn( struct complex Y[][MAX_COLUMNS],
                            struct complex J_vector[],
                            int *num_nodes,
                            char dependentVariable[][5]);

int SetFreq( real freq,
                struct complex Y_of_s[][MAX_COLUMNS],
                int N,
                struct complex Y[][MAX_COLUMNS]);
```

```
int GetBranchData(
                        enum componentTypes *component,
                        char labels[][10],
                        real values[],
                        int *hNode,
                        int *iNode,
                        int *jNode,
                        int *kNode,
                        logical *done,
                        char depVar[],
                        char depVar2[]);

int DecodeUnits( char units[],
                        enum unitTypes *baseUnits,
                        real *multiplier);

int DecodeOnePort(      char token[][10],
                        enum unitTypes expectedUnits,
                        real values[],
                        char labels[][10],
                        int *jNode,
                        int *kNode);

int GaussianElimination(  int max_row,
                               struct complex A[][MAX_COLUMNS],
                               struct complex b[],
                               struct complex x[]);

int DoolittleMethod( int max_row,
                        struct complex A[][MAX_COLUMNS],
                        int d[],
                        struct complex b[]);

int SubstituteLU(       int n,
                        struct complex b[],
                        struct complex LU[][MAX_COLUMNS],
                        struct complex x[],
                        int indexSwapper[]);

/*************************************************/
/*  functions below are for complex arithmetic    */


struct complex cDiv(  struct complex numer,
                        struct complex denom);
```

```
struct complex cMult( struct complex A,
                      struct complex B);

struct complex cSub(  struct complex A,
                      struct complex B);


float arg(struct complex A);


struct complex cmplx( real A,
                      real B);

struct complex cAdd(  struct complex A,
                      struct complex B);



struct complex cSub(  struct complex A,
                      struct complex B);

real cMag(struct complex A);


real cAbs(struct complex A);

struct complex cSqrt( struct complex A);


double cdAbs(struct complex A);

real arg(struct complex A);


struct complex cMult( struct complex A,
                      struct complex B);


struct complex sMult(      real a,
                           struct complex B);

struct complex cDiv(  struct complex numer,
                      struct complex denom);
```

Appendix C

```
/*******************************************************/
/*                                                     */
/*  Appendix C -- Functions for Complex Arithmetic    */
/*                                                     */
/*******************************************************/
#include "globDefs.h"
#include "protos.h"


/***************************************/
/*                                     */
/*     cmplx()                         */
/*                                     */
/*  merges two real into one complex   */
/*                                     */
/***************************************/
struct complex cmplx(  real A, real B)
{
struct complex result;

result.Re = A;
result.Im = B;
return( result);
}

/***************************************/
/*                                     */
/*  cAdd()                             */
/*                                     */
/***************************************/
```

```
struct complex cAdd(
                     struct complex A,
                     struct complex B)
{
struct complex result;

result.Re = A.Re + B.Re;
result.Im = A.Im + B.Im;
return( result);
}
/****************************************/
/*                                      */
/*  cSub()                              */
/*                                      */
/****************************************/
struct complex cSub(
                     struct complex A,
                     struct complex B)
{
struct complex result;

result.Re = A.Re - B.Re;
result.Im = A.Im - B.Im;
return( result);
}
/****************************************/
/*                                      */
/*  cMag()                              */
/*                                      */
/****************************************/
real cMag(struct complex A)
{
real result;
result = sqrt(A.Re*A.Re + A.Im*A.Im);
return( result);
}

/****************************************/
/*                                      */
/*  cAbs()                              */
/*                                      */
/****************************************/
real cAbs(struct complex A)
{
real result;
```

```
result = sqrt(A.Re*A.Re + A.Im*A.Im);
return( result);
}

/****************************************/
/*                                      */
/*  cdAbs()                             */
/*                                      */
/****************************************/
double cdAbs(struct complex A)
{
double result;
result = sqrt(A.Re*A.Re + A.Im*A.Im);
return( result);
}

/****************************************/
/*                                      */
/*  arg()                               */
/*                                      */
/****************************************/
real arg(struct complex A)
{
real result;

if( (A.Re == 0.0)  && (A.Im == 0.0) )
    {
    result = 0.0;
    }
else
    {
    result = atan2( A.Im, A.Re );
    }
return( result);
}

/****************************************/
/*                                      */
/*  cSqrt()                             */
/*                                      */
/****************************************/
struct complex cSqrt( struct complex A)
{
struct complex result;
double r, theta;
```

```
r = sqrt(cdAbs(A));
theta = arg(A)/2.0;
result.Re = r * cos(theta);
result.Im = r * sin(theta);
return( result);
}
/*****************************************/
/*                                       */
/*  cMult()                              */
/*                                       */
/*****************************************/
struct complex cMult(
                        struct complex A,
                        struct complex B)
{
struct complex result;

result.Re = A.Re*B.Re - A.Im*B.Im;
result.Im = A.Re*B.Im + A.Im*B.Re;
return( result);
}
/*****************************************/
/*                                       */
/*  sMult()                              */
/*                                       */
/*****************************************/
struct complex sMult(
                        real a,
                        struct complex B)
{
struct complex result;

result.Re = a*B.Re;
result.Im = a*B.Im;
return( result);
}
/*****************************************/
/*                                       */
/*  cDiv()                               */
/*                                       */
/*****************************************/
struct complex cDiv(
                    struct complex numer,
                    struct complex denom)
```

```
{
real bottom,real_top,imag_top;
struct complex result;

bottom = denom.Re*denom.Re + denom.Im*denom.Im;
real_top = numer.Re*denom.Re + numer.Im*denom.Im;
imag_top = numer.Im*denom.Re - numer.Re*denom.Im;
result.Re = real_top/bottom;
result.Im = imag_top/bottom;
return( result);
}
```

Appendix

D

Laplace Transform

The *Laplace transform* is a technique which is useful for transforming differential equations to algebraic equations that can be more easily manipulated to obtain desired results. The Laplace transform is named for the French mathematician Pierre Simon de Laplace (1749–1827).

In most communications applications, the functions of interest will usually (but not always) be functions of time. The Laplace transform of a time function $x(t)$ is usually denoted as $X(s)$ or $\mathscr{L}[x(t)]$ and is defined by

$$X(s) = \mathscr{L}[x(t)] = \int_0^\infty x(t)e^{-st}\, dt \tag{D.1}$$

The complex variable s is usually referred to as the *complex frequency* and has the form $\sigma + j\omega$, where σ and ω are real variables sometimes referred to as the *neper frequency* and *radian frequency*, respectively. The Laplace transform for a given function $x(t)$ is obtained by simply evaluating the given integral. Some mathematics texts denote the time function with an uppercase letter and the frequency function with a lowercase letter. However, the use of lowercase for time functions is almost universal in the engineering literature.

If we transform both sides of a differential equation in t, using the definition (D.1), we obtain an algebraic equation in s which can be solved for the desired quantity. The solved algebraic equation can then be transformed back to the time domain by using the inverse Laplace transform. The inverse Laplace transform is defined by

$$x(t) = \mathscr{L}^{-1}[X(s)] = \frac{1}{2\pi j} \int_C X(s)e^{st}\, ds \tag{D.2}$$

where C is a contour of integration chosen so as to include all singularities of $X(s)$. The inverse Laplace transform of a given function $X(s)$ can be obtained by evaluating the given integral. However, this integration is often

TABLE D.1 Laplace Transform Pairs

Reference #	$x(t)$	$X(s)$
1	1	$\frac{1}{s}$
2	$u_1(t)$	$\frac{1}{s}$
3	$\delta(t)$	1
4	t	$\frac{1}{s^2}$
5	t^n	$\frac{n!}{s^{n+1}}$
6	$\sin \omega t$	$\frac{\omega}{s^2+\omega^2}$
7	$\cos \omega t$	$\frac{s}{s^2+\omega^2}$
8	e^{-at}	$\frac{1}{s+a}$
9	$e^{-at} \sin \omega t$	$\frac{\omega}{(s+a)^2+\omega^2}$
10	$e^{-at} \cos \omega t$	$\frac{s+a}{(s+a)^2+\omega^2}$

a major chore—when tractable, it usually involves application of the residue theorem from the theory of complex variables. Fortunately, in most cases of practical interest, direct evaluation of (D.1) and (D.2) can be avoided by using some well-known transform pairs listed in Table D.1, along with a number of transform properties presented in Table D.2.

TABLE D.2 Properties of the Laplace Transform

	Property	Time function	Transform
1.	Homogeneity	$af(t)$	$aF(s)$
2.	Additivity	$f(t) + g(t)$	$F(s) + G(s)$
3.	Linearity	$af(t) + bg(t)$	$aF(s) + bG(s)$
4.	First derivative	$\frac{d}{dt} f(t)$	$sF(s) - f(0)$
5.	Second derivative	$\frac{d^2}{dt^2} f(t)$	$sF(s) - sf(0) - \frac{d}{dt} f(0)$
6.	kth derivative	$\frac{d^{(k)}}{dt^k} f(t)$	$s^k F(s) - \sum_{n=0}^{k-1} s^{k-1-n} f^{(n)}(0)$
7.	Integration	$\int_{-\infty}^{t} f(\tau)\, d\tau$	$\frac{F(s)}{s} + \frac{1}{s}\left[\int_{-\infty}^{t} f(\tau)\, d\tau\right]_{t=0}$
		$\int_{0}^{t} f(\tau)\, d\tau$	$\frac{F(s)}{s}$
8.	Frequency shift	$e^{-at} f(t)$	$X(s + a)$
9.	Time shift right	$u_1(t - \tau) f(t - \tau)$	$e^{-\tau s} F(s),\ a > 0$
10.	Time shift left	$f(t + \tau),\ f(t) = 0$ for $0 < t < \tau$	$e^{\tau s} F(s)$
11.	Convolution	$y(t) = \int_{0}^{t} h(t - \tau) x(\tau)\, d\tau$	$Y(s) = H(s)X(s)$
12.	Multiplication	$f(t)g(t)$	$\frac{1}{2\pi j} \int_{C - j\infty}^{C + j\infty} F(s - r) G(r)\, dr$ $\sigma_g < C < \sigma - \sigma_f$

Note: $f^{(k)}(t)$ denotes the kth derivative of $f(t)$ and $f^{(0)}(t) = f(t)$.

Index

Augmented nodal formulation, 147–156

Backward substitution, 111
Bandpass filters, 67–68
Bandstop filters, 68–70
Bessel filters, 61, 99–105
  transfer function of, 99–102
  frequency response, 102–104
  group delay response, 104–105
Boltzmann's constant, 32
Branch admittance, 139
Branches, 19–20
Butterworth filters, 61, 70–83
  impulse response, 77–80
  minimum order, 74
  step response, 81–83
  transfer function, 70–73

Capacitance, 10–11
Capacitive reactance, 12
Capacitive susceptance, 12
Capacitors:
  entry in nodal admittance formulation, 141
  ideal, 10–12
  practical, 16–17
Carrier delay, 60
Cauer synthesis, 169–175
Causality, 41
Cavendish, H., 8
Chebyshev filters, 61, 81–98
  frequency response, 89–90, 95–97
  impulse response, 91–94, 98
  step response, 94–98
  transfer function, 83–89
Chebyshev polynomials, 84
Choleski method, 122
Chords, 131
Column vector, 108
Complex frequency, 223
Complex function, 217–221
Complex type, 3
Conductance, 6
  entry in nodal admittance formulation, 140–141
Constitutive equations, 12
Continued fractions, 53–58
Convolution, 42
Cotree, 131
Cramer's rule, 110
Crout's method, 122–125
Current-controlled current source, 31–32
  entry in augmented nodal admittance formulation, 152
Current-controlled voltage source, 31–32
  entry in augmented nodal admittance formulation, 153
Current source:
  current-controlled, 31–32
  entry in nodal admittance formulation, 141–142
  independent, 8
Cutset matrix, 133–134
Cutsets, 131

Daraf, 12
Dependent node, 185
Dependent sources, 31–32
Dielectric, 9
Diodes, 32–35
Doolittle's method, 122, 125–128
Dynamic conductance, 35

Edge, 129
Elastance, 12
Elements of a matrix, 107
Envelope delay, 60

Factorization of matrices, 122–128
Farad, 10
Faraday, M., 12

Field-effect transistor, 36–37
File naming, 2
Filters:
  bandpass, 67–68
  bandstop, 68–70
  Bessel, 61, 99–105
  Butterworth, 61, 70–83
  Chebyshev, 61, 81–98
  highpass, 67
Flux, 13
Flux linkage, 12–14
Forward elimination, 111
Foster synthesis, 164–169
Fractions, continued, 53–58
Frequency response, 58

Gaussian elimination, 110–122
  with full pivoting, 117–118
  naive, 111–112
  with partial pivoting, 114–116
  practical, 112–118
  with scaled partial pivoting, 116–117
  working position, 112
Graphs, 129–131
Group delay, 59–60

h-parameter model, 35
Heaviside expansion, 52
Henry, J., 14
Highpass filters, 67

Impedances in augmented nodal admittance formulation, 148
Impulse response, 42
  scaling of, 67
Incidence matrix, 132
Incremental network, 200–202
Independent current source, 8
Independent voltage source, 8
Inductance, mutual, 14
Inductive reactance, 14
Inductive susceptance, 14
Inductors:
  entry in nodal admittance formulation, 141
  ideal, 12–14
  practical, 17

Kirchhoff's current law, 27, 135
Kirchhoff's voltage law, 28

Ladder networks, synthesis of, 175–183
  double-terminated *LC*, 181–183
  load-terminated *LC*, 178–180
  source-terminated *LC*, 175–177
Laguerre method, 48
Laplace transform, 223–225
Linear systems, 39–51
Linearity, 40
Logical type, 3
Loops, 131, 185
Loop matrix, 132–133
LU factorization, 122

Magnitude response, 58
Magnitude scaling, 65, 164
Mason's rule, 191–193
Matrix, 107–108
Matrix addition, 108
Matrix equations, 109
Matrix multiplication, 108–109
Mho, 7
Multiplication:
  matrix, 108–109
  scalar, 108
Mutual inductance, 14

Neper frequency, 223
Network configuration, computer specification of, 20–21
Network elements, 5–18
Network theory, 19–30
Networks, ladder, synthesis of, 175–183
  double-terminated *LC*, 181–183
  load-terminated *LC*, 178–180
  source-terminated *LC*, 175–177
Nodal admittance formulation, 135–144
Node voltage, 139
Nodes, 19–20
Norton transformation, 30

Ohm, G. S., 8
Operational amplifier, 155
Operators, 39
Optimization, 1

Parameter extraction method, 197
Partial fraction expansion, 51–52
Passband, 61
Path, 129, 131, 185
Phase delay, 59–60
Phase response, 58
  scaling of, 65
Pivoting strategies, 112–119, 121–122
Poles, 45–47
Power dissipation, 7–8
Prototypes, 213–216

Radian frequency, 223
Reactance:
  capacitive, 12

Reactance (*Cont.*):
inductive, 14
Real type, 3
Reciprocity, 28–29
Resistance, 5–6
Resistors:
ideal, 5
practical, 15–16
standard values for 15–16
Response:
frequency, 58
group delay, 59–60
impulse, 42
scaling of, 67
magnitude, 58
phase, 58
phase delay, 59–60
step, 42–43
scaling of, 67
Row vector, 108

Saturation current, 32
Scalar multiplication, 108
Scaling:
of impulse response, 67
magnitude, 65, 164
of phase response, 65
of step response, 67
Sensitivity, 199–202
analysis via incremental networks, 200–202
relative, 199
Siemens, E. W. von, 8
Siemens (unit), 7
Signal flow graphs, 185–194
Source node, 185
SPICE, 1
Steady-state response, 58
Step response, 42–43
scaling of, 67
Stopband, 61
Subgraph, 129
Superposition, 28–29, 40
Susceptance:
capacitive, 12
inductive, 14
Synthesis:
Foster I, 164–167
Foster II, 167–169
Cauer I, 169–172
Cauer II, 173–175
double-terminated *LC* ladder networks, 181–183
of input impedance, 163–175
insertion loss, 181–183
load-terminated *LC* ladder networks, 178–180
source-terminated *LC* ladder networks, 175–177

Tableau formulation, 144–146
Thevenin transformation, 30
Time-invariance, 40–41
Transfer function, 43–48
conditions for lumped-parameter realization, 44
Transformers, 14–15
entry in augmented nodal admittance formulation, 154–155
Transistors:
bipolar junction, 35, 37
field-effect, 36–37
Transition band, 61
Tree enumeration, 193–196
Trees, 131
Triangular factorization, 122–128
Types, 3
Twigs, 131

Voltage-controlled current source, 31–32, 34
entry in nodal admittance formulation, 142
Voltage-controlled voltage source, 31–32
entry in augmented nodal admittance formulation, 151–152
Voltage divider, 28
Voltage source, independent, 8

Weber (unit), 13
Weber-turn, 13

Zeros, 45–47

ABOUT THE AUTHOR

C. Britton Rorabaugh received his BSEE and MSEE from Drexel University. He is the author of *Signal Processing Design Techniques*, *Data Communications and Local Area Networking*, and *Communications Formulas and Algorithms*, published by McGraw-Hill and TAB/McGraw-Hill.

DISK WARRANTY

This software is protected by both United States copyright law and international copyright treaty provision. You must treat this software just like a book, except that you may copy it into a computer to be used and you may make archival copies of the software for the sole purpose of backing up our software and protecting your investment from loss.

By saying, "just like a book," McGraw-Hill means, for example, that this software may be used by any number of people and may be freely moved from one computer location to another, so long as there is no possibility of its being used at one location or on one computer while it is being used at another. Just as a book cannot be read by two different people in two different places at the same time, neither can the software be used by two different people in two different places at the same time (unless, of course, McGraw-Hill's copyright is being violated).

LIMITED WARRANTY

McGraw-Hill warrants the physical diskette(s) enclosed herein to be free of defects in materials and workmanship for a period of sixty days from the purchase date. If McGraw-Hill receives written notification within the warranty period of defects in materials or workmanship, and such notification is determined by McGraw-Hill to be correct, McGraw-Hill will replace the defective diskett(s). Send requests to:

Customer Service
TAB/McGraw-Hill
13311 Monterey Avenue
Blue Ridge Summit, PA 17294-0850

The entire and exclusive liability and remedy for breach of this Limited Warranty shall be limited to replacement of defective diskette(s) and shall not include or extend to any claim for or right to cover any other damages, including but not limited to, loss of profit, data, or use of the software, or special, incidental, or consequential damages or other similar claims, even if McGraw-Hill has been specifically advised to the possibility of such damages. In no event will McGraw-Hill's liability for any damages to you or any other person ever exceed the lower of suggested list price or actual price paid for the license to use the sofware, regardless of any form of the claim.

McGRAW-HILL, INC. SPECIFICALLY DISCLAIMS ALL OTHER WARRANTY, EXPRESS OR IMPLIED, INCLUDING BUT NOT LIMITED TO, ANY IMPLIED WARRANTY OF MERCHANTABILITY OR FITNESS FOR A PARTICULAR PURPOSE. Specifically, McGraw-Hill makes no representation or warranty that the software is fit for any particular purpose and any implied warranty of merchantability is limited to the sixty-day duration of the Limited Warranty covering the physical diskette(s) only (and not the software) and is otherwise expressly and specifically disclaimed.

This limited warranty gives you specific legal rights; you may have others which may vary from state to state. Some states do no allow the exlusion of incidental or consequential damages, or the limitation on how long an implied warranty lasts, so some of the above may not apply to you.